T0237445

Quantum Computing for the Quantum Curious

Ciaran Hughes • Joshua Isaacson
Anastasia Perry • Ranbel F. Sun
Jessica Turner

Quantum Computing for the Quantum Curious

Springer

Ciaran Hughes
Batavia
IL, USA

Joshua Isaacson
Batavia
IL, USA

Anastasia Perry
Naperville
IL, USA

Ranbel F. Sun
North Reading
MA, USA

Jessica Turner
Batavia
IL, USA

ISBN 978-3-030-61603-8   ISBN 978-3-030-61601-4   (eBook)
https://doi.org/10.1007/978-3-030-61601-4

Acknowledgments

It is a pleasure to thank Marge Bardeen, Harry Cheung, and Spencer Pasero for their helpful discussions on various aspects of this project, from inception to completion. We are grateful to Daniel Carney, William Jay, Yin Lin, Jim Simone, Julia Stadler, Liner de Souza Santos, and Anders Ellers Thomsen for reading and providing feedback on the draft document. It is also a pleasure to thank LaMargo Gill for her remarkably thorough proofreading of this document. We thank Heath O'Connell and Aaron Sauers for their useful advice regarding information content. We thank Olivia Vizcarra and the Fermilab Theory Group for facilitating this project.

This work would not be possible without funding from the Robert Noyce Teacher Scholarship and the Fermilab Teacher Research Associates (TRAC) program. This work was supported by the Fermi Research Alliance, LLC, under Contract No. DE-AC02-07CH11359 with the U.S. Department of Energy, Office of Science, Office of High Energy Physics, and partial support was received by an HEP-QIS QuantISED award titled "Quantum Information Science for Applied Quantum Field Theory." Additionally, we are very appreciative of IBM for financially sponsoring the book, with special thanks to Abraham Asfaw and Sebastian Hassinger, without which this course would not have been open source.

Various sources were used as inspiration for building this course. We acknowledge IBM Q experience[1] for their useful web interface and note that specific figures (as indicated in their captions) are owned by IBM as per their end-user license agreement.[2] We urge the reader to review this end-user license agreement before using the IBM Q web interface. Additionally, we would like to acknowledge the useful PhET Interactive Simulations[3] supplied by the University of Colorado Boulder. Furthermore, we credit the Quantum Mechanics Visualization Project (QuVis),[4] hosted by the University of St. Andrews, for useful interactive simulations. Finally,

---

[1] https://quantum-computing.ibm.com.

[2] https://quantum-computing.ibm.com/terms.

[3] https://phet.colorado.edu.

[4] https://www.st-andrews.ac.uk/physics/quvis/.

we thank Martin Laforest and the Communications and Strategic Initiatives Team at the Institute for Quantum Computing, University of Waterloo's outreach department[5] for supplying material that formed the inspiration for Chaps. 3, 5, and 9 of this module.

[5]https://uwaterloo.ca/institute-for-quantum-computing/outreach.

Contents

Course Description

About

Quantum computing is a growing field at the intersection of physics and computer science. This module introduces three key principles of quantum computing: superposition, quantum measurement, and entanglement. The goal of this course is to bridge the gap between popular science articles and advanced undergraduate texts, making some of the more technical aspects accessible to high school students, early undergraduates, or the scientifically literate general public. Problem sets and simulation-based labs of various levels are included to reinforce the concepts described in the text.

The module starts by covering basic quantum mechanics concepts needed to understand quantum computing. However, it is not designed to be a comprehensive introduction to modern physics. Rather, the course will focus on the topics that students may have heard about but are not typically covered in a typical physics class.

The module is intended to take approximately 15–20 hours to complete. Given the usual constraints on teaching time, these materials could be used after the AP exams, in an extracurricular club, or as an independent project resource to provide students with an overview of quantum computing.

Answers to odd-numbered exercises are included in this book. Answers to even-numbered exercises can be accessed by course instructors at springer.com/10.1007/978-3-030-61601-4.

Prerequisites

The material assumes knowledge of waves from high school physics. Introductory modern physics (photoelectric effect, wave/particle duality, etc.) is helpful but not required, and computer programming experience is not necessary.

The units are labeled by difficulty depending on the level of math and abstract reasoning involved. It is possible to skip over intermediate and/or advanced topics depending on the student's background. For those who are rusty on probability and linear algebra concepts, a refresher is provided in Appendices A and B.

● Fundamental

- Grades: 9–10
- Math Prerequisites: probability of flipping one coin, histograms

■ Intermediate

- Grades: 11–12
- Math Prerequisites: trigonometry, matrix multiplication, probabilities of flipping multiple coins

◆ Advanced

- Grades: 12+
- Math Prerequisites: vectors, vector spaces, matrices as transformations

Learning Objectives

1. Introduction to Superposition

   - Explain what it means for an object to be in a quantum superposition.
   - Identify the measurement outcome of a system in a classical vs. quantum superposition.

   Key Terms: *quantum system, quantum state, quantum superposition*

2. What is a Qubit?

   - Explain the difference between a classical bit and a qubit.
   - Write a mathematical expression for the superposition of a two-state particle using "ket" notation.
   - Compute the probability of finding the particle in a particular state given a normalized superposition state.
   - Express a qubit's state as a vector and/or visually using the Bloch sphere.
   - Perform matrix multiplication to change the qubit's state.

   Key Terms: *qubit, ket notation, state amplitude, normalization, Bloch sphere, unitary matrix*

3. Creating Superposition: The Beam Splitter

   - Explain how light behaves like a particle in the single-photon beam splitter experiment.

- Show how the beam splitter creates a particle in a superposition state.
- Trace the path of light through a Mach–Zehnder interferometer from both a wave interference and a particle perspective.

Key Terms: *photon, beam splitter, phase shift, Mach–Zehnder interferometer*

4. Creating Superposition: Stern–Gerlach

- Explain why electron spin could serve as an example of a qubit.
- Show how the Stern–Gerlach experiment illustrates spin quantization, superposition, and measurement collapse.
- Define what is meant by a measurement basis and convert a given spin to a different basis.
- Compute the probability of an electron passing through one or more Stern-Gerlach apparatuses.

Key Terms: *spin, Stern–Gerlach experiment, measurement basis, orthogonal states, no-cloning theorem*

5. Quantum Cryptography

- Send a message with the one-time pad to understand what is meant by a cryptographic key.
- Generate a shared key using the BB84 quantum key distribution protocol.
- Show how the principles of superposition and measurement collapse make the protocol secure.

Key Terms: *key, quantum key distribution*

6. Quantum Gates

- Build and test quantum circuits on IBM's quantum computer.
- Interpret the histograms produced by single qubit gates: the X, Hadamard, and Z gates.
- Predict the output of multiple gates in a row, including two successive Hadamards.
- Use the matrix representation of gates to determine the new state of the system.

Key Terms: *quantum gates, X gate, Hadamard gate, Z gate*

7. Entanglement

- Show how measurement affects the state of entangled particles.
- Write the state of a multi-qubit system in "ket" notation.
- Identify whether two qubits are entangled given a particular state.

- Predict the output of circuits involving CNOT gates.
- Entangle two qubits using gates.

Key Terms: *quantum entanglement, product/separable states, entangled states, CNOT gate*

8. Quantum Teleportation

- Explain how entanglement is used to transmit the state of a qubit from one place to another.
- Explain the limitations and paradoxes of quantum teleportation.

Key Terms: *quantum teleportation, no-cloning theorem*

9. Quantum Algorithms

- List the benefits and limitations of quantum computers.
- Describe how superposition and interference are leveraged in quantum computing algorithms.

Key Terms: *quantum parallelism, Deutsch–Jozsa algorithm*

Alternative Pathways

The units are best studied in numerical order. However, for those with limited time, Figure 1 shows the minimum recommended prerequisites for each unit. A few references and examples may have to be skipped over, but the core content should still be understandable.

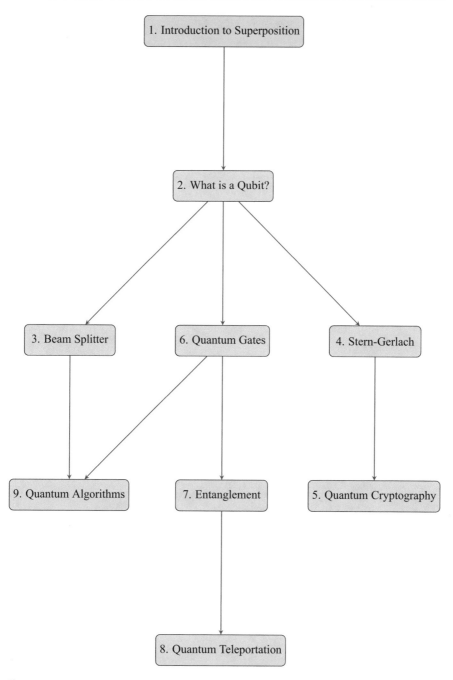

Fig. 1 Flowchart of learning outcomes.

Introduction to Superposition

1

In this section, we review the concepts of classical and quantum superposition. Quantum superposition is the framework for understanding all quantum phenomena. As we do not observe quantum phenomena in our everyday lives, it may seem confusing at first. However, as unintuitive as the quantum world may appear, there are a vast number of experiments which conclusively show that the universe really does operate according to the law of quantum superposition at the smallest distances accessible today.[1] Before going into specific details on quantum superposition, it is useful to explain how the term "superposition" is used in different contexts in both classical and quantum physics. At the end of the chapter, we present the related activities and questions. After gaining experience with quantum superposition from working through these problems, it will become more intuitive. The more experience you gain by advancing through this book, the more quantum superposition will make sense.

1.1 ● Classical Superposition

In classical physics, the concept of **superposition** is used to describe when two physical quantities are added together to make another third physical quantity that is entirely different from the original two. An example of the "superposition principle" in classical physics is clear when working with waves. Two pulses on a string which pass through each other will interfere following the principle of superposition as shown Fig. 1.1. Noise-canceling headphones use superposition by creating sound waves with the same magnitude as the incoming sound wave but completely out of

---

[1]These experiments have culminated in tests of Bell's inequality https://en.wikipedia.org/wiki/Bell_test_experiments —showing that particles can actually be in two locations at the same time https://www.quantamagazine.org/physicists-are-closing-the-bell-test-loophole-20170207/.

© The Author(s) 2021
C. Hughes et al., *Quantum Computing for the Quantum Curious*,
https://doi.org/10.1007/978-3-030-61601-4_1

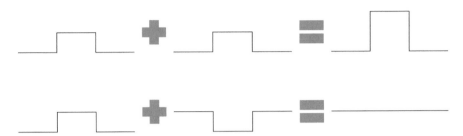

Fig. 1.1 Examples of constructive and destructive interference due to the classical superposition principle

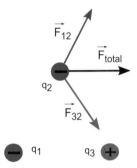

Fig. 1.2 A classical superposition is used to calculate the total electric force on a charge q_2 due to charges q_1 and q_3

phase, thereby canceling the sound wave. This destructive interference is illustrated in the second figure of Fig. 1.1.

Another common application of classical superposition is finding the total magnitude and direction of quantities such as force, electric field, magnetic field, etc. For example, to calculate the total electric force $\vec{F}_{total}$ on a charge q_2 produced by other charges q_1 and q_3, one would sum the forces produced by each individual charge: $\vec{F}_{total} = \vec{F}_{12} + \vec{F}_{32}$. The challenge here is that forces are vectors, so vector addition is needed, as shown in Fig. 1.2.

1.2 ⬤ Quantum Superposition

Quantum superposition is a phenomenon associated with quantum systems. Quantum systems include small objects such as nuclei, electrons, elementary particles, and photons, for which the wave-particle duality and other non-classical effects are observed. For example, you would normally expect that an object can have an arbitrary amount of kinetic energy ranging from 0 to infinity (∞) Joules, i.e. a baseball could be at rest or thrown at any speed. However, according to quantum mechanics, the ball's energy is **quantized**, meaning it can only have certain values.

Classical Systems   Quantum Systems

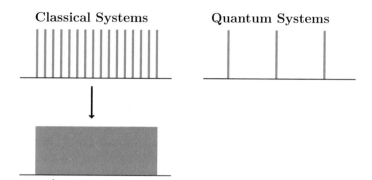

Fig. 1.3 Quantum effects associated with energy quantization are important at the atomic and subatomic distances. In this figure, the grey lines represent allowed energies. In quantum systems, the energies are quantized. As we zoom out of the quantum system to see it through a classical lens (represented by the downward arrow), the energies become more dense and appear continuous. This is the reason quantization is not noticeable in everyday objects

Fig. 1.4 A tossed coin has a 50% chance of landing on heads or tails

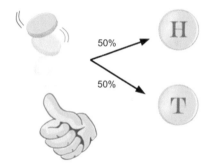

A specific example of energy quantization is when energies can only have integer values $E = 0, 1, 2, 3, \ldots$, but not any numbers inbetween. This is counterintuitive, as we cannot observe it with our classical eyes. The gaps in energy are too small to be seen with the human eye and as such can be treated as continuous for classical physics. However, the gaps are more pronounced at smaller sizes, as shown in Fig. 1.3. For example the hydrogen atom is small enough that quantum effects are important, and Bohr needed to quantize the energy levels to successfully model its behavior.

One aspect of quantum superposition can be explained using a coin analogy. A coin has a 50/50 probability of landing as either heads or tails, as shown in Fig. 1.4.

Question 1 What state is the coin in while it is in the air? Is it heads or tails?

We can say that the coin is in a superposition of both heads and tails. When it lands, it has a **definite state**, either heads or tails. Generally, the word "state" means any particular way that a system can possibly be described. For example, the coin can be either heads, or tails, or a combination of heads or tails while flipped in the air. All of these cases are called states of the coin system. While the coin is being flipped it is in a state of superposition. When we observe the coin, we are making a measurement which destroys the superposition.

At any given time, a system can be described as being in a particular state. The state is related to its quantized values. For example, a tossed coin is either in a heads state or a tails state. An electron orbiting a hydrogen atom could be in the ground state or an excited state. A quantum system is special because it can be in a superposition of these definite states, i.e., both heads and tails simultaneously. The outcome of a measurement is to observe some definite state with a given probability.

In Schrödinger's famous thought experiment, Schrödinger's cat is placed in a closed box with a single atom that has some probability of emitting deadly radiation at any time. Since radioactive nuclear decay is a spontaneous process, it is impossible to predict for certain when the nucleus decays. Therefore, you do not know whether the cat is alive or dead unless you open and look in the box. (Watch this video.)[2] It can be said that the cat is both alive **AND** dead with some non-zero probability. That is, the cat is in a quantum superposition state until you open the box and measure its state. Upon measurement, the cat is obviously either alive **OR** dead and the superposition has collapsed to a definite, non-superposition state.

Quantum systems can exist in a superposition state, and measuring the system will collapse the superposition state into one definite classical state. This might be hard to understand from a classical point of view, as we usually do not see quantum superposition with our human eyes (i.e in macroscopic objects). Einstein was really bothered by this feature of quantum systems. His friend, Abraham Pais, records: "I recall that during one walk, Einstein suddenly stopped, turned to me, and asked whether I really believed that the moon exists only when I look at it."[3]

1.3   Big Ideas

1. A particle in a quantum superposition exists as a combination of different states at the same time.
2. Each possible state has a given probability of being observed, but measurement destroys the superposition because only one definite state is seen.

1.4   Activities

● Quantum Tic-Tac-Toe in Worksheet 10.3

---

[2]https://www.youtube.com/watch?v=uWMTOrux0LM.

[3]Nielsen, M. A. 1., & Chuang, I. L. (2000). Quantum computation and quantum information. New York: Cambridge University Press, p. 212.

Fig. 1.5 Image of the
painted suns

1.5   Check Your Understanding

1. Discuss whether the following quantities are quantized or continuous:
   (a) electric charge
   (b) time
   (c) length
   (d) cash
   (e) paint color
2. ● An ink is created by mixing together 50% red ink and 50% yellow ink. An
   artist uses it to stamp a picture of a sun. If the ink behaves like a quantum system
   in a half-yellow, half-red quantum superposition, what are the different options
   for what the resulting picture could look like? Some options are shown in Fig. 1.5.
3. ● If this controversial picture of a dress[4] is always seen as blue/black by
   Student A and always seen as white/gold by Student B, is the dress in a quantum
   superposition?

What Is a Qubit?

2

In classical computers, information is represented as the binary digits 0 or 1. These are called bits. For example, the number 1 in an 8-bit binary representation is written as 00000001. The number 2 is represented as 00000010. We place extra zeros in front to write every number with 8-bits total, which is called one byte. In fact, every classical computer translates these bits into the human readable information on your electronic device. The document you read or video you watch is encoded in the computer binary language in terms of these 1's and 0's. Computer hardware understands the 1-bit as an electrical current flowing through a wire (in a transistor) while the 0-bit is the absence of an electrical current in a wire. These electrical signals can be thought of as "on" (the 1-bit) or "off" (the 0-bit). Your computer then decodes the classical 1 or 0 bits into words or videos, etc.

Quantum bits or **qubits** are similar to bits in that there are two measurable states called the 0 and 1 states. However, unlike classical bits, qubits can also be in a superposition state of these 0 and 1 states, as shown in Fig. 2.1. Certain computations that would normally need to be performed on 0 or 1 separately on a classical computer could now be completed in a single operation using a qubit on a quantum computer. Intuitively, this could make computations much faster. It is important to understand that although a single qubit is in a superposition of two classical bits, when a qubit is measured, the measurement actually only results in one classical bit of information: either 0 or 1.

2.1 ● Mathematical Representation of Qubits

2.1.1 Dirac Bra-Ket Notation

In order to work with qubits, it is useful to know how one can express quantum mechanical states with mathematical formulas. Dirac or "bra-ket" notation is commonly used in quantum mechanics and quantum computing. The state of a qubit

© The Author(s) 2021
C. Hughes et al., *Quantum Computing for the Quantum Curious*,
https://doi.org/10.1007/978-3-030-61601-4_2

Fig. 2.1 A classical bit can be either 0 or 1. A qubit can be in a superposition of both 0 and 1

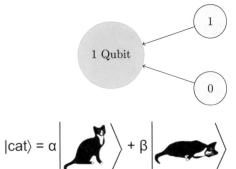

Fig. 2.2 The state of Schrödinger's cat expressed in bra-ket notation

is enclosed in the right half of an angled bracket, called the **"ket"**. A qubit, $|\Psi\rangle$, could be in a $|0\rangle$ or $|1\rangle$ state or even a superposition of both $|0\rangle$ and $|1\rangle$. This is written as

$$|\Psi\rangle = \alpha|0\rangle + \beta|1\rangle, \tag{2.1}$$

with α and β called the amplitudes of the states (Fig. 2.2). Amplitudes are generally complex numbers (a special type of number used in mathematics and physics). However, to understand the meaning of amplitudes, we can imagine the amplitudes as being ordinary (real) numbers. Amplitudes allow us to mathematically represent all of the possible superpositions.

   Amplitudes are very important because they give us the probability of finding the particle in that specific state when performing a measurement. The probability of measuring the particle in state $|0\rangle$ is $|\alpha|^2$, and the probability of measuring the particle in state $|1\rangle$ is $|\beta|^2$. Why is it squared? The short answer is that it gives the correct experimental predictions for this choice of representation.[1] Squaring α and β to find the probability is similar to squaring a wave's amplitude to find the energy of the wave. Since the total probability of observing all the states of the quantum system must add up to 100%, the amplitudes must obey this rule:

$$|\alpha|^2 + |\beta|^2 = 1. \tag{2.2}$$

This is called a **normalization** rule. The coefficients α and β can always be rescaled by some factor to normalize the quantum state.

---

[1] We know that quantum physics is probabilistic from experiments. The squared coefficients are needed to make a quantity that behaves like a probability distribution, i.e., it is a real number and positive. There cannot be a negative probability by definition.

2.1.2 Examples

1. The quantum state of a spinning coin can be written as a superposition of heads and tails. Using heads as $|1\rangle$ and tails as $|0\rangle$, the quantum state of the coin is

$$|\text{coin}\rangle = \frac{1}{\sqrt{2}}\left(|1\rangle + |0\rangle\right). \tag{2.3}$$

What is the probability of getting heads?

The amplitude of $|1\rangle$ is $\beta = 1/\sqrt{2}$, so $|\beta|^2 = \left(1/\sqrt{2}\right)^2 = 1/2$. So the probability is 0.5, or 50%.

2. A weighted coin has twice the probability of landing on heads vs. tails. What is the state of the coin in "ket" notation?

$$P_{\text{heads}} + P_{\text{tails}} = 1 \quad \text{(Normalization Condition)}$$

$$P_{\text{heads}} = 2P_{\text{tails}} \quad \text{(Statement in Example)}$$

$$\rightarrow P_{\text{tails}} = \frac{1}{3} = \alpha^2$$

$$\rightarrow P_{\text{heads}} = \frac{2}{3} = \beta^2 \tag{2.4}$$

$$\rightarrow \alpha = \sqrt{\frac{1}{3}}, \ \beta = \sqrt{\frac{2}{3}}$$

$$\rightarrow |\text{coin}\rangle = \sqrt{\frac{1}{3}}|0\rangle + \sqrt{\frac{2}{3}}|1\rangle.$$

One common misconception is that the measurement of a single qubit will result in a weighted average of the $|0\rangle$ and $|1\rangle$ states. It is important to note that after you perform the measurement on a single qubit, the qubit is no longer in a superposition but takes on a definite state of either $|0\rangle$ or $|1\rangle$.[2] This means that you would not be able to find α or β from a single qubit. Instead, we need to create many qubits which are in the same quantum state, and then measure how many of the qubits collapse into $|0\rangle$ (giving α) and how many collapse into $|1\rangle$ (giving β). Therefore, multiple identical particles are needed in order to count how many collapse into $|0\rangle$ or $|1\rangle$.

---

[2]When formulating the mathematical representation of quantum mechanics, this is one of four fundamental assumptions that need to be made. The reason for the collapse is still unknown: https://en.wikipedia.org/wiki/Wave_function_collapse. Read more at this link: https://www.quantamagazine.org/how-quantum-trajectory-theory-lets-physicists-understand-whats-going-on-during-wave-function-collapse-20190703/.

2.2   ◆ Matrix Representation

When writing a single qubit in a superposition $|\psi\rangle = \alpha|0\rangle + \beta|1\rangle$, it is useful to use matrix algebra. In matrix representation, a qubit is written as a two-dimensional vector where the amplitudes are the components of the vector

$$|\psi\rangle = \begin{pmatrix} \alpha \\ \beta \end{pmatrix}. \tag{2.5}$$

The states $|0\rangle$ and $|1\rangle$ are usually represented as

$$|0\rangle = \begin{pmatrix} 1 \\ 0 \end{pmatrix}, \quad |1\rangle = \begin{pmatrix} 0 \\ 1 \end{pmatrix}. \tag{2.6}$$

Experimentally, a qubit's state can be changed through some physical action such as applying an electromagnetic laser or passing it through an optical device. Changing a qubit's state through a physical action mathematically corresponds to multiplying the qubit vector $|\psi\rangle$ by some **unitary matrix** U so that after the operation the state is now $|\psi'\rangle = U|\psi\rangle$. Unitary is a mathematical term which expresses that U can only act on the qubit in such a way that the total probability $|\alpha|^2 + |\beta|^2$ does not change. A matrix U is unitary if the matrix product of U and its conjugate transpose $U^\dagger$ (called U-dagger) multiply to give the identity matrix: $UU^\dagger = U^\dagger U = \mathbb{1}$. This is very important because, in all mathematical constructions of quantum mechanics, one fundamental assumption is that each (matrix) operator must be unitary. This ensures that after changing any state through an action, the total probability to observe all possible states will still add up to 100%. If this did not happen, then we could not interpret the results of quantum mechanics to be probabilistic, and the results would disagree with the many experiments that have been performed to date. The physical action of interacting with the state corresponds mathematically to applying a unitary operator.

2.2.1   Examples

1. What is the conjugate transpose of the following matrix?

$$A = \begin{pmatrix} 1 & i \\ 1 & i \end{pmatrix}. \tag{2.7}$$

The conjugate transpose of a matrix is found using the following two steps. First, we "conjugate" the complex numbers. The conjugate of a complex number is found by switching the sign of the imaginary part. The complex conjugate of 1 is just 1, while the complex conjugate of $+i$ is $-i$. Second, we transpose the conjugated matrix. Transposing a matrix switches rows with columns, i.e., the first row turns into the first column, second row turns into the second column, etc. Therefore,

$$A^\dagger = \begin{pmatrix} 1 & 1 \\ -i & -i \end{pmatrix}. \tag{2.8}$$

2. Is the above matrix A unitary?

$$AA^\dagger = \begin{pmatrix} 1 & i \\ 1 & i \end{pmatrix} \begin{pmatrix} 1 & 1 \\ -i & -i \end{pmatrix} \tag{2.9}$$

$$= 2 \begin{pmatrix} 1 & 1 \\ 1 & 1 \end{pmatrix} \neq \begin{pmatrix} 1 & 0 \\ 0 & 1 \end{pmatrix}. \tag{2.10}$$

Multiplying A by its conjugate transpose does not produce the identity matrix, so A is not unitary.

3. What is the result of applying the unitary operator X onto a $|0\rangle$ state qubit?

$$X = \begin{pmatrix} 0 & 1 \\ 1 & 0 \end{pmatrix}, \qquad |0\rangle = \begin{pmatrix} 1 \\ 0 \end{pmatrix}. \tag{2.11}$$

$$X|0\rangle = \begin{pmatrix} 0 & 1 \\ 1 & 0 \end{pmatrix} \begin{pmatrix} 1 \\ 0 \end{pmatrix} = \begin{pmatrix} 0 \\ 1 \end{pmatrix} = |1\rangle. \tag{2.12}$$

The X matrix changes the $|0\rangle$ qubit state to the $|1\rangle$ qubit state.

2.3 ■ Bloch Sphere

A single qubit can be visualized using the Bloch sphere. The Bloch sphere is a visual representation of a qubit with similar geometric properties to the unit circle from trigonometry. Each point on the Bloch sphere corresponds to a different possible superposition of a single qubit. The top and bottom of the sphere correspond to the two measurable states of the qubit, $|0\rangle$ and $|1\rangle$. An arrow on the Bloch sphere, which can point to any of the different locations on the surface of the sphere, indicates the current state of the qubit. Figure 2.3 shows four examples of how the Bloch sphere can be used to visualize different qubit states. When the arrow is not pointing directly to the top or bottom of the sphere, the qubit is in a superposition state. For example, everywhere around the equator the qubit has a 50/50 chance of collapsing into $|0\rangle$ or $|1\rangle$ upon measurement. The exact location on the equator corresponds to a distinct state, where the amplitudes can have different signs and be either real or imaginary numbers.

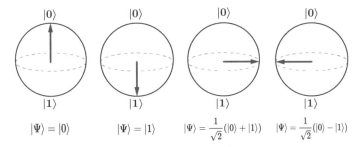

$$|\Psi\rangle = |0\rangle \qquad |\Psi\rangle = |1\rangle \qquad |\Psi\rangle = \frac{1}{\sqrt{2}}(|0\rangle + |1\rangle) \qquad |\Psi\rangle = \frac{1}{\sqrt{2}}(|0\rangle - |1\rangle)$$

Fig. 2.3 The state of a qubit is represented by an arrow on the Bloch sphere

Fig. 2.4 A cartoon of the Bloch sphere depicted as the Earth, and the state of Schrödinger's cat represented as a location on Earth

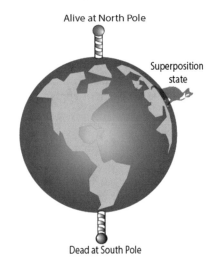

When the state of the qubit is changed, the arrow rotates to a different position on the sphere. One analogy is to think of the qubit like Schrödinger's cat traveling the globe shown in Fig. 2.4. When the cat is at the North Pole, it will definitely be alive. When the cat is at the South Pole, it will definitely be dead. As long as the cat's state is not measured, it can be anywhere else on the globe in a superposition state of alive and dead. As coders of the quantum computer, it is our job to manipulate the state of the qubit which gives the cat instructions on how to move around the globe.

Question 1 Schrödinger's cat is determined to be alive. What location on the Earth in Fig. 2.4 could the cat have been before the quantum measurement?

(a) Russia
(b) Australia
(c) North Pole
(d) all of the above

The cat could have been anywhere on Earth except for the South Pole. Notice that in Australia the cat has a smaller probability of being alive since it is further away from the North Pole.

The Bloch sphere is a helpful visual aide for understanding how a qubit can have an infinite number of possible quantum states. However, it only represents one qubit and does not work for systems of two or more qubits.

2.4 ● Physical Realization of Qubits

In a classical computer, the 0- and 1-bit mathematically represent the two allowed voltages across a wire in a classical circuit. Semiconductor devices called transistors are used to control what happens to these voltages. A question frequently posed by new students is "What is a qubit made out of?" As quantum computers are based on fundamentally different concepts than classical computers, they must be built from completely different technology, i.e. it is not possible to have a classical current in a superposition of both flowing and not flowing through a wire. Quantum computers are still in their infancy, and so there are many different candidates for the technology to build them. Some technologies are based on optical systems, others use superconductors,[3] and there are others based on molecules. It is still unclear if any of these are more beneficial than the others, and it is even more unclear if all future quantum computers will be built from the same technology or if there will be many different types of quantum computers available (in the same way there exists both XBox and PlayStation game consoles, but both have the same general purpose—interactive gaming). We will study two different experiments which illustrate the properties of the qubits, but the engineering details of building a quantum computer are well beyond the scope of this introduction.

2.5 Big Ideas

1. A qubit can be in a superposition of $|0\rangle$ and $|1\rangle$ states. The Bloch sphere can be used to visually represent a single qubit.
2. A qubit can be written in terms of amplitudes. Each squared amplitude corresponds to the probability of measuring the qubit in $|0\rangle$ or $|1\rangle$.
3. A physical change to a qubit mathematically corresponds to unitary matrices which multiply the qubit amplitudes.

---

[3]Fermi National Accelerator Laboratory is researching how to make long-lived coherent qubits using their superconducting radio-frequency cavity expertise, i.e., https://qis.fnal.gov/superconducting-quantum-systems/.

Table 2.1 Table for message

Character	Binary code	Character	Binary code
A	01000001	N	01001110
B	01000010	O	01001111
C	01000011	P	01010000
D	01000100	Q	01010001
E	01000101	R	01010010
F	01000110	S	01010011
G	01000111	T	01010100
H	01001000	U	01010101
I	01001001	V	01010110
J	01001010	W	01010111
K	01001011	X	01011000
L	01001100	Y	01011001
M	01001101	Z	01011010

2.6   Check Your Understanding

1. ⬤ If a coin is a classical bit of information (heads = 1 and tails = 0), how is the number 2 represented in standard 8-bit notation using coins? (Hint: Find the 8-bit representation of the number 2, then convert to H's and T's.)

2. ⬤ Using Table 2.1, can you figure out what this binary message 01000011 01000001 01010100 says? (Note: This is actually how your computer and phone decode information from bits to text.)

3. ⬤ Assume a flipped coin can be measured as either heads (H) or tails (T).
   (a) If the coin is in a normalized state $\frac{1}{\sqrt{10}}|H\rangle + \frac{3}{\sqrt{10}}|T\rangle$, what is the probability that the coin will be tails?
   (b) During a flip, the coin is in a state $\frac{1}{3}|H\rangle + \frac{2}{3}|T\rangle$. Is this state normalized?
   (c) A machine is built to flip coins and put them into a state $\frac{1}{2}|H\rangle + \frac{\sqrt{3}}{2}|T\rangle$ when flipped. If 100 coins are flipped, how many coins should land on tails?
   (d) A coin starts in the state $\frac{1}{\sqrt{10}}|H\rangle + \frac{3}{\sqrt{10}}|T\rangle$. After a measurement is made on the coin, what could be the state of the coin?

4. ⬤ Your friend gives you many qubits which are in same superposition state. How can you determine what the state is?

5. ⬤ A qubit is prepared in an unknown state. It is then measured with the outcome $|0\rangle$.

(a) Which of the following could be its initial state before the measurement:
$|0\rangle$, $\frac{1}{\sqrt{10}}|0\rangle + \frac{3}{\sqrt{10}}|1\rangle$, $\frac{1}{2}|0\rangle + \frac{\sqrt{3}}{2}|1\rangle$ and/or $\frac{1}{\sqrt{2}}(|0\rangle + |1\rangle)$?

(b) If you tried to measure the same qubit a second time, can you narrow down what the initial state was?

(c) Another qubit is prepared in the same unknown state. It is measured in the $|1\rangle$ state. What can you say about the initial state now?

6. ■ What is the matrix product of the X matrix,

$$X = \begin{pmatrix} 0 & 1 \\ 1 & 0 \end{pmatrix}, \tag{2.13}$$

and the $|0\rangle$ state qubit?

7. ■ What is the matrix product of the above X matrix and the $|1\rangle$ state qubit?

8. ■ What is the matrix product of the above X matrix and a qubit in the general state $|\Psi\rangle = \alpha|0\rangle + \beta|1\rangle$?

9. ◆ Find the conjugate transpose of the matrix

$$Y = \begin{pmatrix} 0 & -i \\ i & 0 \end{pmatrix}. \tag{2.14}$$

10. ◆ Show that the matrix

$$U = \frac{1}{\sqrt{2}} \begin{pmatrix} 1 & 1 \\ 1 & -1 \end{pmatrix} \tag{2.15}$$

is unitary.

11. ◆ Show by example that applying a non-unitary matrix to a qubit results in probabilities that no longer add up to 100%. (Hint: Start with any initial state, e.g., $|0\rangle$. Measure the probabilities of finding either 0 or 1. Apply a non-unitary matrix to the initial state. Then measure the probabilities of finding either a 0 or 1. Do the probabilities add up to 100%?)

12. ■ If the qubit represented by Fig. 2.5 is measured, what are the possible outcomes? Numerical values for the amplitudes are not needed, only conceptual statements.

Fig. 2.5 A qubit's state is shown on the Bloch sphere

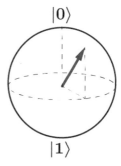

Creating Superposition: The Beam Splitter

3

Now that we have explored qubits and the phenomenon of superposition, we can ask the question: how do we know that superposition actually happens? What is the evidence that shows that a quantum particle really does exist in two different locations at this same time while in a quantum superposition? The nature of science means that experiments are constantly updating previous results, so are there other interpretations of the experimental results that can explain the data without the need for superposition? In this chapter we'll explore the experimental evidence that debunks interpretations other than quantum superposition. Further, while a flipping coin is a simple model of a qubit, it is not very useful for building a quantum computer because it does not exhibit all of the properties of a true quantum superposition. For example, we cannot manipulate the superposition amplitudes. In this chapter, we will study some real physical examples of quantum particles in a superposition containing two states. These examples include a photon in a beam splitter and the Mach–Zehnder interferometer.

3.1 ● Beam Splitter

In classical optics, a **beam splitter** acts like a partially reflective mirror that splits a beam of light into two. In a 50/50 beam splitter, 50% of the light intensity is transmitted and 50% is reflected, as shown in Fig. 3.1.

One way to visualize the beam splitter is to imagine a barrier with holes randomly cut out like Swiss cheese, as shown in Fig. 3.2. Imagine this barrier is placed in a pond, and a water wave moves toward the barrier. After the wave hits the barrier, we would observe a smaller wave going through the barrier and another would be reflected off the barrier.

© The Author(s) 2021

C. Hughes et al., *Quantum Computing for the Quantum Curious*,

https://doi.org/10.1007/978-3-030-61601-4_3

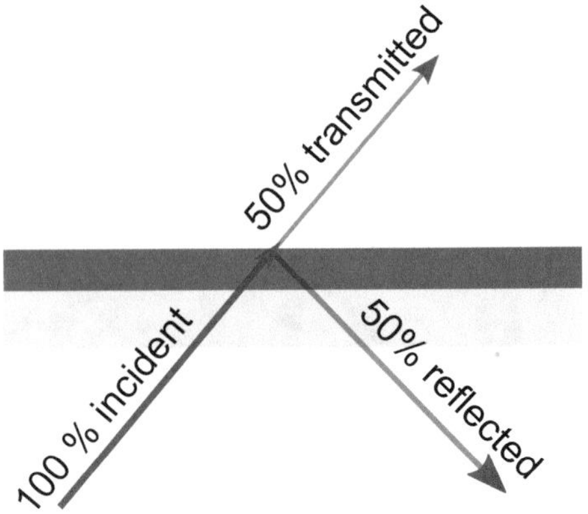

Fig. 3.1 A beam splitter reflects 50% of the incident light and transmits 50% of the incident light.

Fig. 3.2 A beam splitter reflects 50% of the incident light and transmits 50% of the incident light.

Fig. 3.3 Low-intensity light is a stream of single photons.

Question 1 What would happen if a classical particle such as a soccer ball is randomly kicked at the barrier? Assume the ball can fit through the holes.

Experiments demonstrate that light behaves both like a wave (Young's double-slit experiment) and a particle (photoelectric effect, Compton effect). Classically, light is thought of as a wave consisting of continually oscillating electric and magnetic fields. However, light can also be thought of as a stream of particles called **photons**. Photons have no mass but carry the light's energy from one point to another at the speed of light. A laser beam is comprised of photons. If you turn down the intensity of your laser, you can even send one photon at a time, as shown in Fig. 3.3. As setting up a single photon source and detector requires specialized equipment, we will instead run a simulator to explore the quantum effects of photons.

Question 2 Open the beam splitter simulator,[1] go to the Controls screen, and fire a single photon. The setup before the photon hits a beam splitter is shown in Fig. 3.4. Which detectors are triggered when the photon passes through the 50/50 beam splitter?

(a) Always detector 1
(b) Always detector 2
(c) Detector 1 OR detector 2
(d) Both detector 1 AND detector 2
(e) Neither

Question 3 Which detector(s) would trigger if a classical **wave** is sent through the beam splitter?

(a) Always detector 1
(b) Always detector 2
(c) Detector 1 OR detector 2
(d) Both detector 1 AND detector 2
(e) Neither

Question 4 Which detector(s) would trigger if a classical **particle** is sent through the beam splitter?

(a) Always detector 1
(b) Always detector 2

Fig. 3.4 A single photon is sent at a beam splitter and the outcome is measured with detectors to see whether the beam splitter transmits or reflects.

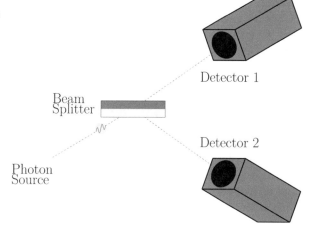

[1] https://www.st-andrews.ac.uk/physics/quvis/simulations_html5/sims/photons-particles-waves/photons-particles-waves.html.

(c) Detector 1 OR detector 2
(d) Both detector 1 AND detector 2
(e) Neither

Question 5 What does the photon do at the instance it encounters the 50/50 beam splitter?

(a) Splits in half. Half the photon is transmitted and half is reflected
(b) The whole photon goes through with 50% probability and reflects with 50% probability
(c) The whole photon is both transmitted and reflected, essentially in two places at once

If the photon was split in half, both detectors in the beam splitter experiment would be triggered at the same time. As only one detector goes off at a time, the photon could not have split up. In this case, we see that light behaves more like the soccer ball than the water wave.

At this point you may be thinking that the photon was either transmitted or reflected at the beam splitter, and we simply didn't have that information until it hit Detector 1 or 2. Unfortunately, this would be the incorrect interpretation formed by our classical animal brain. This would be like saying the coin was Heads all along, and all we had to do was look at it to determine its state. Similarly to how a spinning coin will land on heads 50% of the time and tails 50% of the time, the single photon is in a superposition of both states all the way until the point when it reaches the detectors. This distinction might seem like a matter of semantics, but this is important as the distinction describes two different ways that the universe operates at the smallest possible distances. Also, it will be important once the system becomes more complicated. The experimental setup after the photon hits a beam splitter is shown in Fig. 3.5.

If we let the transmitted path be $|0\rangle$ (detector 1), and the reflected path be $|1\rangle$ (detector 2), then the photon's state after the beam splitter is

$$|\text{photon}\rangle = \frac{1}{\sqrt{2}}|0\rangle + \frac{1}{\sqrt{2}}|1\rangle. \tag{3.1}$$

Upon measurement, will the superposition collapse into either $|0\rangle$ or $|1\rangle$? Unfortunately, it is not possible to predict which detector will be activated at any given time as quantum mechanics is inherently probabilistic.

The phenomenon of superposition allows quantum computers to perform operations on two bits of information at once with a single qubit. In fact, it is possible to create a general purpose (also called universal) quantum computer using photons as qubits, beam splitters to create superposition, and pieces of glass that slow down the photons along selected paths (phase shifters).[2]

---

[2] Knill, E.; Laflamme, R.; Milburn, G. J. (2001). "A scheme for efficient quantum computation with linear optics". Nature. Nature Publishing Group. 409 (6816): 46–52.

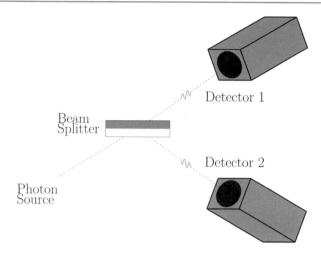

Fig. 3.5 The beam splitter puts the photon into a superposition state.

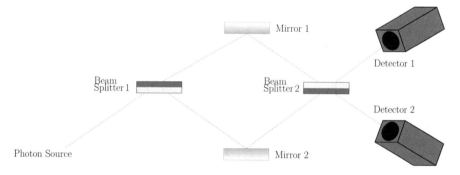

Fig. 3.6 Schematic   of   the   Mach–Zehnder   interferometer   from   https://www.st-andrews.ac.uk/physics/quvis/simulations_html5/sims/Mach-Zehnder-Interferometer/Mach_Zehnder_Interferometer.html

3.2 ■ Mach–Zehnder Interferometer

To convince ourselves that the photon really did take two paths at once, let's see what happens when a second beam splitter is added. This experimental setup is shown in Fig. 3.6. The mirrors redirect the photons towards the second beam splitter. This device configuration is known as a **Mach–Zehnder interferometer**. The set up is very sensitive to the distances between the mirrors and detectors, which have to be the same or differ by an integer number of the photon's wavelength.

Question 6 If we assume that the photon was reflected by the first beam splitter, which detectors would be triggered?

(a) Always detector 1
(b) Always detector 2

(c) Detector 1 OR detector 2
(d) Both detector 1 AND detector 2
(e) Neither

Question 7 If we assume that the photon was transmitted by the first beam splitter, which detectors would be triggered?

(a) Always detector 1
(b) Always detector 2
(c) Detector 1 OR detector 2
(d) Both detector 1 AND detector 2
(e) Neither.

Question 8 Construct the Mach–Zehnder interferometer in the beam splitter simulator[3] and fire a single photon. Which detectors are triggered?

(a) Always detector 1
(b) Always detector 2
(c) Detector 1 OR detector 2
(d) Both detector 1 AND detector 2
(e) Neither

If the photon was either transmitted or reflected by the first beam splitter, it would have a 50/50 chance of transmission or reflection by the second beam splitter. Thus, both detectors should trigger with equal probability. However, strangely the experimental results do not agree with this hypothesis, as only one detector is triggered with 100% probability. This weird phenomenon is more intuitively understood from the wave perspective of light.

To understand the operation of the interferometer, it is important to note that the beam splitters have a polarity. The beam splitter consists of a piece of glass coated with a dielectric on one side. When light enters the beam splitter from the dielectric side, the reflected light is **phase shifted** by π. Light entering from the glass side will not experience any phase shift. The phase shift only occurs when the light travels from a low to high index of refraction ($n_{air} < n_{dielectric} < n_{glass}$).

What does it mean for a photon to be phase shifted? In this case, it is more intuitive to think about the wave nature of light. The phase shift would invert the electric and magnetic field oscillations relative to the incoming wave. If a π-shifted wave overlaps with the original wave, destructive interference occurs as is shown in Fig. 3.7.

---

[3]https://www.st-andrews.ac.uk/physics/quvis/simulations_html5/sims/Mach-Zehnder-Interferometer/Mach_Zehnder_Interferometer.html.

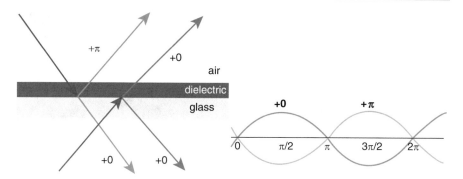

Fig. 3.7 The light through a beam splitter is phase shifted if it is reflected from the dielectric side but not phase shifted if it is reflected from the glass side.

Question 9 If we assume that light is a classical wave exhibiting interference, can you work out which detectors would be triggered? Note that the first beam splitter has the dielectric side on top, while the second has the dielectric on the bottom, as shown in Fig. 3.6.

(a) Always detector 1
(b) Always detector 2
(c) Detector 1 OR detector 2
(d) Both detector 1 AND detector 2
(e) Neither

3.2.1 Particle Explanation

The behavior of the interferometer can also be viewed from the particle perspective, though it may be less intuitive. Recall from the single beam splitter experiment that the photon did not split up or clone itself. It was in a superposition state, essentially taking both paths. The second beam splitter treats the photon as if it came in from both top and bottom simultaneously. As shown in Fig. 3.8, the top path enters the second beam splitter from the glass side and experiences no phase shift, whereas the bottom path enters from the dielectric side and is phase shifted upon reflection. The $+0$ and $+\pi$ states at Detector 2 interfere destructively, while the $+0$ and $+0$ states at Detector 1 interfere constructively. Therefore, Detector 1 triggers with 100% probability.

Question 10 If the photon is sent into the Mach–Zehnder interferometer from the upper left instead of the bottom left, which detector(s) would be triggered and with what probability?

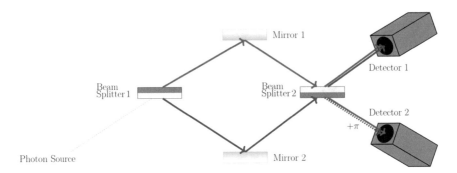

Fig. 3.8 The blue path shows the photon's path if it is reflected by Beam Splitter 1. The red path shows the path if the photon is transmitted. Because Beam Splitter 2 has the dielectric facing downwards, blue is phase shifted upon reflection.

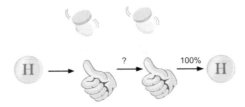

Fig. 3.9 Coin analogy for the interferometer. Sending a photon through one beam splitter puts it in superposition, but adding a second beam splitter undoes the superposition and recovers the original state.

Even though the output of the first beam splitter is 50/50, the second beam splitter can distinguish whether the laser was fired from the top or the bottom. The first beam splitter creates a superposition state, but adding a second one undoes the superposition and recovers the original state. This is a non-classical operation. It would be like starting with the coin heads up, flipping it, flipping it again while it is still in the air, and then always getting heads when it lands! This is highlighted in Fig. 3.9.

There is hidden information in the superposition state. In the Mach–Zehnder photon qubit, the information is encoded in the form of the phase shift. In the experiment shown in Fig. 3.8, we chose the phase shift to have a value of π. However, we could have just as easily chosen the phase shift to have any value between 0 and 2π (the angles of a circle). Each separate choice of phase shift would produce a different type of superposition state that would still produce the same measurable 50/50 outcome. This is represented on the Bloch sphere by different locations along the equator.[4] This phase shift information is present in

---

[4]A complex amplitude $e^{i\phi}$ with infinite possible phase angles ϕ does not affect the probability since $|e^{i\phi}|^2 = 1$.

the amplitudes but not the square of the amplitudes (and hence hidden from us in the Mach–Zehnder experiment–though we could make another experiment to try to determine this information). Here are two simple examples of distinct states that can be created in two different experimental arrangements of the Mach–Zehnder experiment which still have the same 50/50 probability:

$$\frac{1}{\sqrt{2}}|0\rangle + \frac{1}{\sqrt{2}}|1\rangle \quad \text{or} \quad \frac{1}{\sqrt{2}}|0\rangle - \frac{1}{\sqrt{2}}|1\rangle. \tag{3.2}$$

In these two states the plus or minus signs represents two of the many different phase shifts that are possible. Each different choice of the phase shift depends on how the experimental arrangement is chosen. As you can see, quantum superposition is inextricably linked to wave-particle duality.

Furthermore, in the Mach–Zehnder experiment we created a superposition, performed a phase shift and then observed wave interference. These experimental operations are equivalent to mathematically applying matrix/gate operations on a qubit, as we shall see later. As such, the Mach–Zehnder is an example of how we can technologically implement qubits (the photon) and operations (superposition/phase shift, etc) to build a quantum computer.[5] In quantum computing, people talk about the superposition of states rather than the wave behavior. Yet, as we have seen, both frameworks lead to the same understanding of the Mach–Zehnder interferometer. Later we will use the interferometer to implement a quantum algorithm.

3.3   Big Ideas

1. A photon can be put into a superposition using a beam splitter. After passing through the beam splitter, a photon takes both paths simultaneously.
2. The Mach–Zehnder interferometer shows how the photon really does take two paths at once. This is conclusive experimental evidence of superposition of photons.

3.4   Check Your Understanding

1. ● Your friend who is explaining superposition to you says that:
   "A particle in the state $(1/\sqrt{2})|0\rangle + (1/\sqrt{2})|1\rangle$ represents a lack of knowledge of the system. Over time, the particle is changing back and forth between the state $|0\rangle$ and $|1\rangle$. The superposition state says that overall, the particle is in each of the two states for half of the time."
   What parts of this statement do you agree with and what do you not agree with?

---

[5]It should be noted that the technology has progressed so that most qubits are at present implemented using superconducting transmons and not using a Mach–Zehnder.

2. ■ Only one detector is triggered if a single photon is sent through the beam splitter experiment shown in Fig. 3.5. If the laser outputs two photons at the same time, what is the probability that both detectors will be triggered simultaneously? Now how about three photons? Ten photons? Note that this is why a higher power beam of light appears to reach both detectors simultaneously.

3. ◆ In practice, it is difficult to place the detectors the exact same distance from the beam splitter. The difference in distance is measured using the time delay Δt between photons. The experiment is shown in Fig. 3.10 and the data in Fig. 3.11.

   (a) Does the data shown in Fig. 3.11 at $\Delta t = 0$ support that light is a particle or a wave?

Fig. 3.10 The experiment varies the position of Detector 2 and records the number of coincidences, i.e., the number of times both detectors are triggered simultaneously.

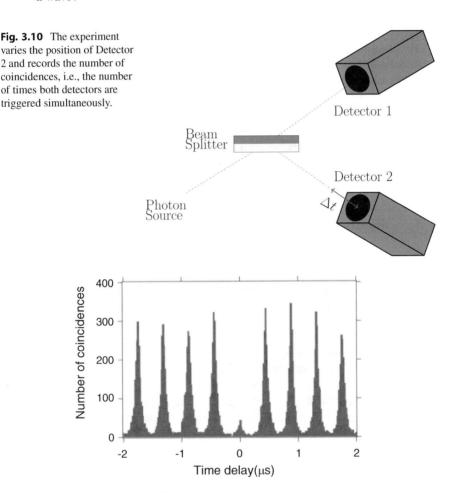

Fig. 3.11 Data is shown above for light bursts sent from the laser every 0.4 μs. Figure reproduced with permission of Martin Laforest and the Communications and Strategic Initiatives Team at the Institute for Quantum Computing, University of Waterloo Outreach department.

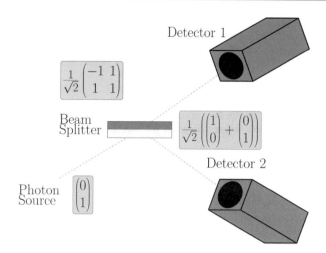

Fig. 3.12 Matrix formulation of the Mach–Zehnder apparatus.

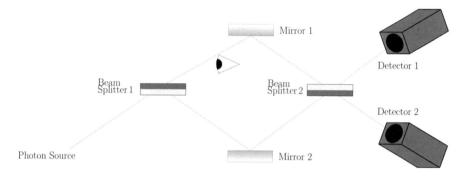

Fig. 3.13 A third detector (your eye) is added to the Mach–Zehnder apparatus.

(b) Why are there large coincidence counts when $\Delta t \neq 0$? (Hint: Look at the spacing between the peaks.)

4. ■ Using the matrices given in Fig. 3.12, show how the superposition state is created by multiplying the beam splitter matrix by initial photon state.

5. ◆ Construct the matrix representation for a 30/70 beam splitter.

6. ■ Unsettled by the Mach–Zehnder interferometer, you decide to determine once and for all which path the photon takes after the first beam splitter. You place another detector (indicated by the eyeball) on the upper path as shown in Fig. 3.13. If the eyeball sees a photon, what would be seen at Detectors 1 and 2?

Creating Superposition: Stern–Gerlach

4

In the previous chapter, we have seen that a photon in an interferometer can be a prototype for a qubit. Might there be any other prototypes for a qubit arising from other particles that we might know? In fact, an electron is another prototype for a qubit. An electron has many measurable properties such as energy, mass, momentum. But, for the purposes of creating a qubit, we want to focus on a property with only two measurable values. An electron has a two-state property which is called **spin**.

Classically, an electron's spin can be visualized as a rotation about its own axis, like a spinning top or fidget spinner. You learned in high school physics that a moving charge creates a magnetic field according to the right-hand rule. By curling the fingers of your right hand in the direction of the electron's rotation, your thumb points in the direction of the magnetic field created by the charge. Conceptually, an electron's spin behaves somewhat like a tiny bar magnet. However, this classical picture is just an analogy. In reality, the quantum mechanical property we call "spin" is intrinsic to the electron (like its mass or charge). The property was called spin because it can be described mathematically just like orbital momentum, but spin does not actually correspond to the electron physically rotating.[1] Just like a lot of quantum phenomena, spin can be confusing at first. Exploring how the electron can be used as a qubit will provide further intuition into quantum phenomena such as quantum superposition, spin, and measurement.

4.1 ● Stern–Gerlach Apparatus

The **Stern–Gerlach apparatus** (SGA) showed that the electron spin is quantized to only two values. This video[2] explains the experimental apparatus used to measure the electron's spin. The key point here is that the vertically oriented apparatus (called

---

[1] See https://en.wikipedia.org/wiki/Spin_(physics) for more details.

[2] https://www.youtube.com/watch?v=rg4Fnag4V-E.

© The Author(s) 2021
C. Hughes et al., *Quantum Computing for the Quantum Curious*,
https://doi.org/10.1007/978-3-030-61601-4_4

the z-direction by convention) only measures the spin as either up or down, not randomly oriented at any angle. Since the spin of an electron has two measurable states, it can represent a qubit with $|0\rangle$ as spin up and $|1\rangle$ as spin down (Fig. 4.1).

Question 1 Open up the PhET Stern–Gerlach simulator[3] and try sending electrons of various initial spins into the Stern–Gerlach apparatus (SGA).

Are the results what you would expect? The "up" and "down" directions are defined by the orientation of the apparatus, as in Fig. 4.2. There is nothing inherently special about the z-direction compared to the x- or y-direction. An SGA rotated horizontally would measure either spin left or spin right. An SGA rotated by 45° would measure the spin to be either diagonally up or diagonally down. What is particularly interesting is if we send a single spin up electron into a horizontally oriented SGA.

Question 2 Where would you expect a spin up electron to land in Fig. 4.3 after passing through a horizontal SGA?

Classically, vertically oriented bar magnets in a horizontal magnetic field would land at the center of the screen. However, recall that the spin can only be measured as left or right and cannot possibly land in the center. The way quantum mechanics solves this problem is to have the electron land either on the left or the right with 50% probability. Sound familiar? Sending a spin up electron through a horizontal SGA puts the electron in a superposition state of left and right.

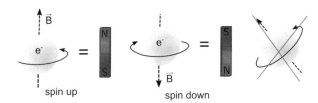

Fig. 4.1 An electron can spin either up or down and produce a magnetic field.

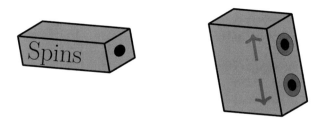

Fig. 4.2 A cartoon picture of the Stern–Gerlach Apparatus. Electron spin produces a magnetic field either in the up or down direction.

---

[3]https://phet.colorado.edu/sims/stern-gerlach/stern-gerlach_en.html.

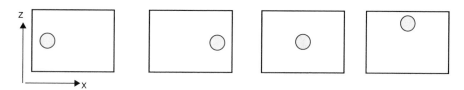

Fig. 4.3 Choices for Question 2.

The Stern–Gerlach experiment shows that qubits in superposition are an accurate description of how nature truly operates. Therefore, one promising application of quantum computers is simulating systems that occur in nature such as electronic properties of a molecule for use in drug design.[4]

4.2 ◆ Measurement Basis

Spin in the vertical direction can be represented as a superposition of spins in the horizontal direction. As shown in the simulation, an electron with vertical spin has a 50% chance of being measured as right or left:

$$|\uparrow\rangle = \frac{1}{\sqrt{2}}|\rightarrow\rangle + \frac{1}{\sqrt{2}}|\leftarrow\rangle, \tag{4.1}$$

$$|\downarrow\rangle = \frac{1}{\sqrt{2}}|\leftarrow\rangle - \frac{1}{\sqrt{2}}|\rightarrow\rangle. \tag{4.2}$$

In more traditional qubit notation, spin in the $+z$ and $-z$ axis is written as $|0\rangle$ and $|1\rangle$, while spin in the $+x$ and $-x$ axis is $|+\rangle$ and $|-\rangle$:

$$|0\rangle = \frac{1}{\sqrt{2}}|+\rangle + \frac{1}{\sqrt{2}}|-\rangle, \tag{4.3}$$

$$|1\rangle = \frac{1}{\sqrt{2}}|+\rangle - \frac{1}{\sqrt{2}}|-\rangle. \tag{4.4}$$

This is non-classical because you cannot add or subtract horizontal magnetic field vectors to get a vertical magnetic field vector. One analogy might be to think about a person looking at a coin vertically to determine its state. If they see heads or tails, someone looking from the side would see a superposition. If they are forced to make a choice via measurement, they would say heads or tails with 50% probability (Fig. 4.4).

---

[4]https://analyticsindiamag.com/top-applications-of-quantum-computing-everyone-should-know-about/.

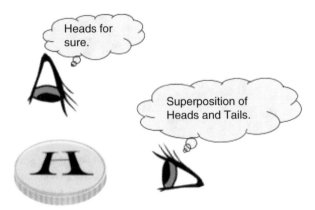

Fig. 4.4 Analogy for how a definite vertical spin is seen as a superposition in the horizontal direction.

Example Write the $|+\rangle$ state in terms of $|0\rangle$ and $|1\rangle$.

Solution Adding Eqs. (4.3) and (4.4) we find

$$|0\rangle + |1\rangle = \frac{2}{\sqrt{2}}|+\rangle. \tag{4.5}$$

Rearranging, we get

$$|+\rangle = \frac{1}{\sqrt{2}}|0\rangle + \frac{1}{\sqrt{2}}|1\rangle. \tag{4.6}$$

Similarly, by subtracting Eqs. (4.3) and (4.4), we find

$$|-\rangle = \frac{1}{\sqrt{2}}|0\rangle - \frac{1}{\sqrt{2}}|1\rangle. \tag{4.7}$$

These equations show that a horizontal spin is a superposition of spin up and spin down. As we saw in the beam splitter example, the minus sign encodes information about the original state of the particle before it is put in superposition. As described and visually shown in Sect. 2.3, it is possible to choose other complex amplitudes that give the same probability, but the details are mathematically beyond our scope.

We reached the conclusion that spins in one direction can be written as a superposition of spins in another direction. Within the quantum computing field, the "z-basis" is composed of $|0\rangle$ and $|1\rangle$, while $|+\rangle$ and $|-\rangle$ compose the "x-basis." A basis is analogous to a coordinate system for quantum states. Any state can be

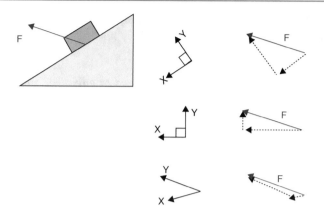

Fig. 4.5 Rewriting quantum states in terms of a different basis is similar to decomposing a classical vector into a different choice of coordinate system.

written in terms of a different choice of basis, similarly to how any vector can be broken down into components along a different choice of axes.

In Fig. 4.5, a box on a ramp is subject to a force. The vector decomposition of $\vec{F}$ is shown for three different coordinate systems. All three coordinate systems are valid for describing the force, but only the first two are convenient to use in actual calculations. By choosing x–y to be perpendicular, you have made the components mutually exclusive: if a vector is horizontal, you know it's definitely not vertical. The x- and y- directions can be treated as two independent problems. The mathematical term for expressing that the axes are independent is "orthogonal". In quantum mechanics, there are an infinite number of possible choices for a basis. However, the basis should have two properties:[5]

1. The basis must describe all possible quantum states for the system.
2. The basis must be orthogonal.

Let us check these conditions for the z-basis, which consists of states $|0\rangle$ and $|1\rangle$:

1. Because the Stern–Gerlach experiment shows that an electron is either spin up or spin down, the most general state of the electron would be a superposition of up and down:

$$|\text{electron}\rangle = \alpha|0\rangle + \beta|1\rangle. \tag{4.8}$$

---

[5]These two properties can also be used to form a basis in a classical system, where states should be swapped for vectors.

A linear combination of $|0\rangle$ and $|1\rangle$ completely describes the electron's state.

2. If you measure the spin as $|0\rangle$, it is definitely not $|1\rangle$, therefore $|0\rangle$ and $|1\rangle$ are orthogonal.

The same argument can be made for the x-basis or any other angle of the SGA (Fig. 4.5).

4.3 ◆ Geometric Representation of a Basis

In this geometric representation of the z-basis and x-basis, the orthogonal states are drawn perpendicular to one another. If the electron is in a particular state $|0\rangle$ in the z-basis, the state vector can be decomposed into $1/\sqrt{2}|+\rangle + 1/\sqrt{2}\rangle|-\rangle$ in the x-basis. Physically turning the SGA from vertical to horizontal corresponds to changing the measurement from the z to the x-basis. Since $|0\rangle = 1/\sqrt{2}|-\rangle + 1/\sqrt{2}\rangle|-\rangle$, the spin up particle became a 50/50 superposition when the measurement device became horizontal.

Question 3 Use Fig. 4.6 and trigonometry to show that $|1\rangle = 1/\sqrt{2}|+\rangle - 1/\sqrt{2}\rangle|-\rangle$.

Often, there is hidden information about the state that cannot be measured unless we change to a different basis. In the x-basis, there is no measurable difference between $|0\rangle$ and $|1\rangle$. Both the $|0\rangle$ qubit and the $|1\rangle$ qubit would have measurement results of 50% left and 50% right in the x-basis. In the z-basis, $|0\rangle$ would have 100% probability of being measured up in the Stern–Gerlach and 0% being measured down, while $|1\rangle$ would have 0% probability being measured up and 100% down.

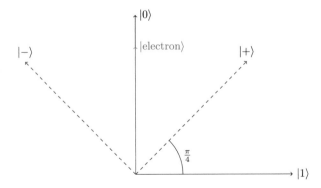

Fig. 4.6 Geometric representation of the z-basis and x-basis. The state of a spin up electron is shown.

4.4 ● **Effect of Measurement**

You learned that measuring a qubit collapses its superposition state into one of two possibilities. A spinning coin is in a superposition state, but once it lands, it becomes either heads or tails. The photon is in a superposition state after passing through a beam splitter, but once it reaches the detectors, we know for sure whether it was reflected or transmitted. To appreciate the truly strange nature of quantum measurement, let's see what happens when electrons are sent through multiple Stern–Gerlach devices in a row.

Question 4 Open the PhET Stern–Gerlach simulator[6] and send electrons with randomly oriented spins through a vertical SGA as in Fig. 4.7. What is the spin of the electrons that pass through the hole?

(a) $+z$
(b) $-z$
(c) Superposition of $+z$ and $-z$

Question 5 Add a second SGA, oriented horizontally as in Fig. 4.8. What is the spin of the electrons before entering the second SGA?

(a) $+x$
(b) $-x$
(c) Superposition of $+x$ and $-x$

Question 6 What is the spin of the electrons after passing through the second SGA?

(a) $+x$
(b) $-x$
(c) Superposition of $+x$ and $-x$

Question 7 What is the z-spin of the electron coming out of the second SGA? Design an experiment to confirm this in the simulation.

Fig. 4.7 The z-axis SGA lets through spin up electrons but blocks spin down electrons.

---

[6]https://phet.colorado.edu/sims/stern-gerlach/stern-gerlach_en.html.

Fig. 4.8 The z and x-axis SGA.

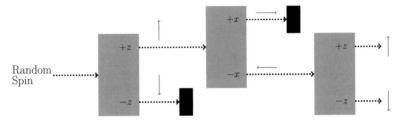

Fig. 4.9 The first SGA selects for $+z$ spin and the second SGA selects for $-x$. The third SGA shows that by measuring the $-x$ in the z-basis then the electron is in a superposition of $+z$ and $-z$.

(a) $+z$
(b) $-z$
(c) Superposition of $+z$ and $-z$

Given that only spin up electrons passed through the first SGA, one would expect that the electron is still spin up after the second SGA, no matter what is measured in x. However, if you measure the z-spin with a third SGA as in Fig. 4.9, it has a 50% chance of being up or down!

By measuring the electron, we fundamentally changed its state. Measuring the x-spin of the qubit puts it into a superposition of up and down, even when it started as up to begin with. When you measure the length of an object with a ruler, you don't expect the object's length to change after you measure the it!

Quantum measurement collapse is used in many quantum applications such as cryptography, where one could detect if a message has been intercepted. This will be discussed in further detail in Chap. 5. Moreover, this property of quantum states implies that a qubit in an unknown state cannot be copied. This concept is known as the no-cloning theorem and has very important consequences. For example, classical computers can make a copy of lines of text and the original version of the text stays the same—there are now two identical copies of the same text. But, if you try to copy an unknown qubit you first have to measure it, which fundamentally alters it by collapsing its superposition state into a basis state. Therefore, since quantum computers cannot copy text as easily as classical computers can, they are unlikely to replace your laptop. However, for certain applications, the information in superposition states allows information processing beyond what is possible in a classical computer. This will be explored more in Chap. 9.

4.5   Big Ideas

1. An electron has an intrinsic property called spin, which is quantized into two values called spin-up and spin-down.
2. The measurement basis is important when interpreting results from experiments on quantum states. Two common basis are the z-basis ($|0\rangle$ and $|1\rangle$) and the x-basis ($|+\rangle$ and $|-\rangle$).
3. The Stern–Gerlach apparatus (SGA) can be used to put the electron into a superposition state. The electron can be used as a qubit, and the SGA as a way to operate on this qubit. Together, they are a simple model of a quantum computer.

4.6   Activities

■ Polarizer Demo in Worksheet 10.2
■ Measurement Basis Lab in Worksheet 10.6
◆ Superposition vs. Mixed States Lab in Worksheet 10.5

4.7   Check Your Understanding

1. ● The Stern–Gerlach apparatus is rotated by 90° so that the magnetic field is in the x-direction as shown in Fig. 4.10. If electrons from a random source are sent through the apparatus, what pattern would be formed on the screen?
2. ◆ Would $|0\rangle$ and $|+\rangle$ together satisfy the criteria for a valid basis?
3. ◆ An electron is in a superposition state shown in the geometric representation in Fig. 4.11.
   (a) What is the state of the electron in the z-basis? i.e. find α and β in $|\text{electron}\rangle = \alpha|0\rangle + \beta|1\rangle$
   (b) What is the probability of measuring spin up?
   (c) What is the state of the electron in the x-basis? i.e, find α and β in $|\text{electron}\rangle = \alpha|+\rangle + \beta|-\rangle$.
   (d) What is the probability of measuring the spin in the $|-\rangle$ direction?
4. ◆ To measure the difference between an electron in a spin state $\frac{1}{\sqrt{2}}|0\rangle + \frac{1}{\sqrt{2}}|1\rangle$ and one in $\frac{1}{\sqrt{2}}|0\rangle - \frac{1}{\sqrt{2}}|1\rangle$, one could use:

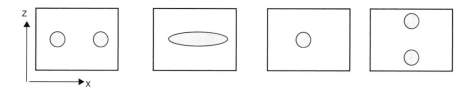

Fig. 4.10 Stern Gerlach apparatus.

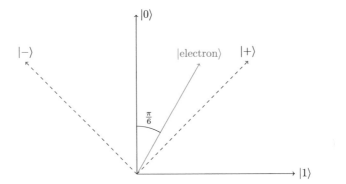

Fig. 4.11 Superposition state of the electron.

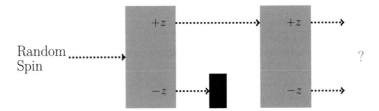

Fig. 4.12 SGA setup for Problem 5.

(I) A horizontal SGA.

(II) A vertical SGA.

(III) A 45° diagonal SGA.

(a) I only
(b) II only
(c) I or III
(d) II or III
(e) I, II, or III

5. ⬤ An electron with random spin is sent through two vertical SGAs as shown in Fig. 4.12. What would be the output of the second SGA?

6. ⬤ An electron with random spin is sent through two vertical SGAs, where the second SGA is rotated upside down, or 180°.
   (a) If the second +z port is blocked as in Fig. 4.13, what would be the output of the second SGA?
   (b) If both ports on the second SGA are open as in Fig. 4.14, what would you see at the output?

7. ⬤ An electron with random spin is sent through a horizontal SGA followed by a vertical SGA as in Fig. 4.15. What would be the output of the second SGA?

8. ⬤ An electron with random spin is sent through three SGAs as shown in Fig. 4.16. What would be the output of the third SGA?

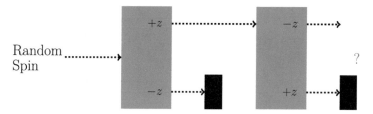

Fig. 4.13 SGA setup for Problem 6a.

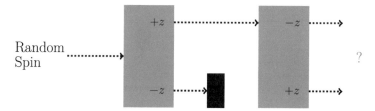

Fig. 4.14 SGA setup for Problem 6b.

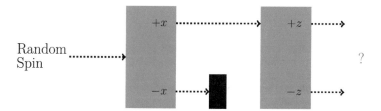

Fig. 4.15 SGA setup for Problem 7.

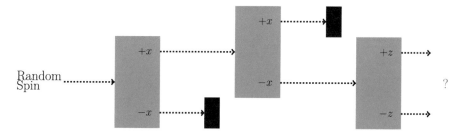

Fig. 4.16 SGA setup for Problem 8.

9. ● An electron with random spin is sent through three SGAs as shown in Fig. 4.17. What would be the output of the third SGA?

10. ● An electron with random spin is sent through four SGAs as shown in Fig. 4.18. What would be the output of the fourth SGA?

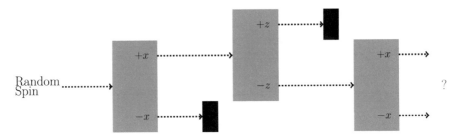

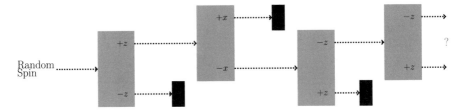

Fig. 4.17 SGA setup for Problem 9.

Fig. 4.18 SGA setup for Problem 10.

Quantum Cryptography

5

The Internet can be thought of as a channel of information being sent from you to everyone else connected to the Internet. If you wanted to transmit your sensitive information (such as bank account numbers or military secrets) over the Internet, then you have to ensure that only the persons you intend to read your information have access to your sensitive data. Otherwise, everyone would be able to read your information, e.g., access to your bank account details and transfer money out of your account. Therefore, one needs to encrypt any data sent over the Internet. Encryption, in this context, ensures that only the intended sender and receiver can understand any message being sent over an Internet channel.

5.1 ● Cryptography Fundamentals

Encryption relies on the sender and receiver sharing a secret key (that no one else has) and using that to encrypt and decrypt messages. In this way, since no one else has the secret key, no one else can understand the shared information. Because no one else understands the shared information, they cannot misuse it for their own benefit.

The only type of encryption protocol known to be perfectly secure is the One-Time Pad, also known as the Vernam Cipher.[1] It is assumed that two people exchange a shared key at least as long as the message in a completely secure way. The shared key encrypts the message to create the cipher, and the cipher is decoded by decrypting with the shared key. The protocol is best understood by trying it out with the associated worksheets in Sect. 10.7. In practice, due to not having a secure channel to share such a complicated key, despite being unbreakable, this method

---

[1] Shannon, Claude (1949). "Communication Theory of Secrecy Systems." *Bell System Technical Journal*. 28 (4): 656–715. https://doi.org/10.1002/j.1538-7305.1949.tb00928.x.

© The Author(s) 2021
C. Hughes et al., *Quantum Computing for the Quantum Curious*,
https://doi.org/10.1007/978-3-030-61601-4_5

is usually not employed.[2] Here we see the fundamental caveat with encryption: you require a secure channel to share the secret key (if you do not have a secure channel then someone random can just take the secret key and encryption would be pointless), but if you have a secure channel then why do you need to encrypt your data? You need a way around this issue. How do you share a secret key in an insecure channel, where anyone can be listening?

5.2 ● Classical Cryptography

The way around sharing a secret key in an insecure channel in the majority of online communications is called public key cryptography.[3] A person called Alice makes two keys such that each key knows that only the other key is related to it (think of the keys as siblings). They are called the private and public key. Alice then gives the public key to everyone in the world but importantly keeps the private key for herself. Anybody else, say Bob, who wants to send a private message to Alice has to encrypt their message with the public key that Alice generated. There are many different types of encryption protocols that one can use. The special part of public key cryptography is that *only* Alice's private key can decrypt the message that was encrypted using its sibling public key. In this way, only Alice can read the message from Bob. As no one else has Alice's private key, no one else can read Bob's message. However, if Bob did not use Alice's public key but used a different public key to encrypt his message, then Alice cannot decrypt that message, as her private key is not a sibling key of the different public key. This whole cryptography scheme relies on the fact that no one can break the encryption protocol. If they could break it, then they could read Alice's message even if they did not have Alice's private key.

The most commonly used modern Internet encryption protocol is RSA encryption. RSA encryption relies on encrypting messages with keys that are made out of very large integers. To break the encryption protocol, an eavesdropper would need to factorize this very large integer into its (prime) factors. Factorizing a large integer into its (prime) factors is known to be a problem that classical computers cannot solve in any reasonable amount of time.[4] For example, given two large prime numbers p and q, it takes just a fraction of a second to multiply these two prime numbers together to produce a large integer $c = pq$. However, finding the two prime numbers p and q given just the integer c would take a classical supercomputer thousands of years.

RSA encryption works by encrypting the message with the public key. Decrypting the message by brute force requires factorizing a large integer in the public

---

[2]https://en.wikipedia.org/wiki/One-time_pad.

[3]https://en.wikipedia.org/wiki/Public-key_cryptography.

[4]https://en.wikipedia.org/wiki/Integer_factorization.

key, which would take thousands of years. However, the private key related to the public key knows how to check the prime factors of the public key and can decrypt the message easily. Because the encryption protocol is so difficult to break, no one would even attempt to do so. Instead, the eavesdropper may attempt to steal your private key by hacking into your computer, which Internet firewalls protect against. As such, nearly all Internet encryption relies on a computer not being able to factor large integers in a short amount of time.

However, in 1995, Peter Shor proposed a quantum computing algorithm, based on superposition and interference, that drastically speeds up the factoring process. A 4000-digit number, which would take a classical computer longer than the lifetime of the universe to factorize, would take less than a day on a large, stable quantum computer. Shor's Algorithm[5] can theoretically break modern encryption schemes, although quantum hardware is not sufficiently advanced yet to make this decryption practical. If it were, all your bank details, military secrets, and industrial secrets could be easily hacked. The details of Shor's algorithm are beyond our scope, so we will instead discuss how the same quantum computer could be used to ensure a key is shared over a secure channel.

Together, the one-time pad and **quantum key distribution** (QKD) would be a formidable combination. The BB84 QKD[6] simulation demonstrates how one could create a shared key using electrons and a Stern–Gerlach apparatus. The BB84 protocol is summarized below.

5.3 ● BB84 Quantum Key Distribution

5.3.1 Before Sending the Message

The sender (Alice) and receiver (Bob) publicly agree to the relationship between spins and bit value shown in Table 5.1.

Table 5.1 Table for the relationship between spin and bit values for quantum cryptography

Spin	↑	←	↓	→
Bit value	0	0	1	1

---

[5]https://quantum-computing.ibm.com/docs/guide/q-algos/shor-s-algorithm.

[6]https://www.st-andrews.ac.uk/physics/quvis/simulations_html5/sims/cryptography-bb84/Quantum_Cryptography.html.

5.3.2 Quantum Part

1. Alice randomly chooses either the x- or z-basis (horizontal or vertical Stern–Gerlach apparatus).
2. Alice sends an electron in superposition in the chosen basis through the SGA, measures the spin, and records the corresponding bit value as 0 or 1. The electron is sent to Bob.
3. Bob randomly chooses either the x- or z-basis.
4. Bob measures the spin of the electron and records whether it was 0 or 1.
5. Repeat steps 1–4 until desired level of security is achieved.

5.3.3 Example

Alice sends five electrons to Bob. When Alice sends an electron prepared in one basis and Bob measures in the same basis, they measure the same spin. However, if Bob measures in a different basis than Alice, then the electron will be in a superposition state and there will be a 50% probability of the state collapsing into 0 or 1. Example values for the first three bits of a BB84 experiment are shown in Fig. 5.1. Can you fill in the last two bits?

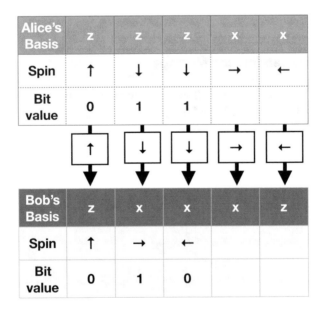

Alice's Basis	z	z	z	x	x
Spin	↑	↓	↓	→	←
Bit value	0	1	1		

Bob's Basis	z	x	x	x	z
Spin	↑	→	←		
Bit value	0	1	0		

Fig. 5.1 Alice's and Bob's measurements of the BB84 protocol

Alice's Basis	z	z	z	x	x
Spin	↑	↓	↓	→	←
Bit value	0	1	1	1	0

Bob's Basis	z	x	x	x	z
Spin	↑	→	←	→	↓
Bit value	0	1	0	1	1

Key	0			1	

Fig. 5.2 Alice and Bob's measurements of the BB84 protocol completed from Fig. 5.1. The discarded bits are grayed out, and the key is 01

5.3.4 Classical Post-processing

1. Alice and Bob publicly share the basis used for each bit measurement *without revealing the actual bit value they measured.*
2. If they measured in the same basis, they keep that bit. If they measured in a different basis, they discard that bit. This is shown in Fig. 5.2. For the measurements performed in the same basis, Alice and Bob are guaranteed to have the same string of bits *unless there was an eavesdropper.*
3. They publicly compare a subset of the bits, say 20 out of 100 bits. If all 20 are the same, then it is unlikely that there was an eavesdropper. The remaining 80 become the shared key.

5.4 ● Detecting an Eavesdropper

If an eavesdropper (Eve) overhears the post-processing part where Alice and Bob share the basis used for each bit measurement, Eve has no information about whether any bit was either a 0 or 1. As Eve has no information, public post-processing sharing is not a dangerous action for Alice and Bob to take. The only way for Eve to determine the spin value of the qubits, and as a consequence acquire important information, is to measure the qubit with her own Stern–Gerlach *before* it gets to Bob. This can be potentially dangerous for Alice and Bob. However, as the basis is not shared during the transmission, Eve must randomly pick a basis to measure the qubit intercepted from Alice. If Alice and Bob randomly choose to measure in a different basis, they throw away all the bits and it does not matter which basis Eve chooses. If Alice and Bob randomly choose to measure in the same basis then there are two outcomes depending on what Eve does: (1) If Eve randomly chooses the same basis as Alice, then she does not alter the state. This is bad, as Eve has successfully eavesdropped information without Alice and Bob knowing. (2) If Eve randomly chooses a different basis than Alice, then she alters the state and puts it into a superposition. Even though Bob is using the same basis as Alice, due to Eve altering the state, Alice and Bob can have a different spin measurement. This is how they can catch an eavesdropper.

5.4.1   Example

The eavesdropping situation is shown in Fig. 5.3. If Eve chooses the same basis as Alice, the spin is unchanged when it gets to Bob (bit #1). If Eve chooses a different basis than Alice, the spin could be different when it gets to Bob (bits #2 and #3). Eve could get lucky and Bob's bit could agree with Alice (bit #2). However, Bob is equally likely to measure something different from Alice (bit #3). Can you fill in what might happen with bits #4 and #5?

When Alice and Bob compare a portion of their key bits, a discrepancy would indicate the presence of an eavesdropper. If they compare a sufficient number of key bits and all of them match, they can be reasonably sure that the rest of it is secure. This statement will be quantified shortly in the questions.

5.5   Big Ideas

1. Classical RSA encryption assumes that factoring a large integer into its prime factors is prohibitively difficult. This assumption is true for classical computers, ensuring your information can be safe.
2. Shor's algorithm on a large and stable quantum computer could factor a large integer into prime factors, making classical encryption vulnerable.
3. New quantum encryption protocols are developed to keep information safe in the quantum era. The BB84 protocol is one way to share a secret key in a secure channel, that can then be used for encryption.

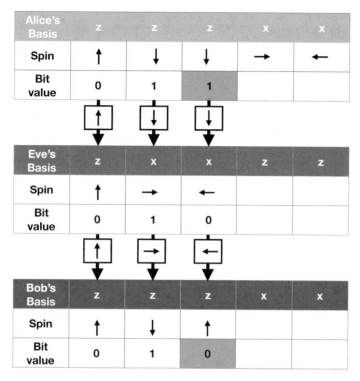

Fig. 5.3 An example of how to catch an eavesdropper using the BB84 protocol

5.6   Activities

- One-time Pad for Alice/Bob in Worksheet 10.7
- BB84 Quantum Key Distribution for Alice/Bob/Eve in Worksheet 10.8
- For those interested in hands-on experiments, see QuTools[7]

5.7   Check Your Understanding

1. ■ If Alice and Bob exchange 1 million bits in order to use the BB84 quantum cryptography protocol, approximately how long will their bit-key string be? Assume they do not check for eavesdropping.
2. ■ Alice and Bob share their lists of measurement basis, but do not share any more information about the bits. What is the probability that Eve will guess the correct bit for a single bit-key?

---

[7]https://www.qutools.com/quantenkoffer_science-kit/.

3. ■ Alice and Bob perform 20 bit-key measurements but do not share any information about the bits. What is the probability that Eve will guess the correct 20-bit key?

4. ● If Eve tries all possible key combinations with the one-time pad, can she crack the one-time pad?

5. ■ If Eve uses a Stern–Gerlach to measure the spin in between Alice and Bob's measurements, what percentage of the time will she be lucky and get the correct key-bit value without detection?

6. ■ If Alice and Bob measure in the same basis and compare 20 bits of their key, what is the probability that Eve could have eavesdropped all 20 bits without being detected?

7. ● Suppose that Eve discovers that the no-cloning theorem is wrong and finds a way to clone the state of each photon. How could she use a cloning machine to learn about the entire key without leaving any trace?

Quantum Gates

6

As discussed in Chap. 2, information in classical computers is represented by bits. However, if the bits did not change, then the computer would remain the same forever and would not be very useful! Therefore, it is necessary to change the values of bits depending on what you want the computer to do. For example, if you want a computer to multiply the number 2 and the number 3 together to produce the number 6, then you need to put each of the numbers 2 and 3 into an 8-bit binary representation, and then have a computational operation to multiply the two 8-bit values together to produce 6. The operation of changing bits in a classical computer is performed by classical logic gates.

6.1   ● Single Qubit Gates

Classical computers manipulate bits using classical logic gates such as OR, AND, NOT and NAND. This link[1] provides a basic review of classical logic gates. Similarly, quantum computers manipulate qubits using quantum gates. The gates are applied to qubits and the states of the qubits change depending on which gate is applied. In the Bloch sphere representation, the gate provides instructions for rotating the qubit's arrow around the sphere. A quantum algorithm has to be implemented on a quantum computer using quantum gates. After running a quantum algorithm, the result is retrieved by measuring the qubit's state. The hardware implementation of quantum gates depends on how the qubit and quantum computer has been implemented technologically.[2] As an example, one could have a qubit based on spin. Then gates could be implemented using an external magnetic field to

---

[1] https://whatis.techtarget.com/definition/logic-gate-AND-OR-XOR-NOT-NAND-NOR-and-XNOR.

[2] E.g., topological qubits and superconducting qubits have very different hardware implementations due to their very different nature.

© The Author(s) 2021
C. Hughes et al., *Quantum Computing for the Quantum Curious*,
https://doi.org/10.1007/978-3-030-61601-4_6

change the spin and hence the qubit state. This chapter will focus on gates from the computing perspective rather than the engineering perspective. You will learn about several important gates that act on a single qubit, interpret histograms produced by a quantum computer simulator, and use matrices to describe the operation of these gates.

6.2 X (Also Called NOT) Gate

In classical computers, the NOT gate takes one input and reverses its value. For example, it changes the 0 bit to a 1 bit or changes a 1 bit to a 0 bit. This is like a light-switch flipping a light from ON to OFF, or from OFF to ON. A quantum X gate is similar in that a qubit in a definite state $|0\rangle$ will become $|1\rangle$ and vice versa. When the qubit is in a superposition of all basis states, then the superposition also flips:

$$\alpha|0\rangle + \beta|1\rangle \quad \boxed{X} \quad \beta|0\rangle + \alpha|1\rangle \tag{6.1}$$

To see how this works, you can try out the IBM Q simulator.[3] Traditionally, all qubits on the IBM Q machine (or any other quantum simulator) start with the incoming qubits in the $|0\rangle$ state. To run this simple gate, drag the X gate onto any qubit. To find the results, add the measurement operation at the end, as shown in Fig. 6.1. Figure 6.1 is known as a *quantum circuit*, the quantum analog to classical circuits. A circuit describes how a qubit changes through a computation depending on which gates act on it. The circuit is read from left to right. As an example, in Fig. 6.1 the single qubit on the left is initialized to $|0\rangle$. An X gate is then applied to that specific qubit, and the last symbol on the qubit line denotes that the qubit is measured. The double line underneath is used to illustrate the measurement.

After running the quantum circuit and opening the results, you should see a histogram showing the measurements of the qubit's final state for 1,024 independent trial runs. As the qubit always starts as the $|0\rangle$ state, applying the X gate produces the $|1\rangle$ state and so the measurement outcome is $|1\rangle$ 100% of the time as shown in Fig. 6.2.

Fig. 6.1 Applying the X gate on the IBM Q simulator and measuring the output

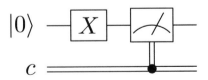

---

[3] https://quantum-computing.ibm.com. It can also be run on IBM's real quantum computer, but you may have to wait in a queue for the results.

Mathematically, the quantum NOT gate is represented as a matrix X which acts on qubit states using matrix multiplication. The matrix representation is

$$X = \begin{pmatrix} 0 & 1 \\ 1 & 0 \end{pmatrix}. \tag{6.2}$$

It is worth noting that any computer will have hardware errors. In a classical computer, this could be an electrical short of the motherboard, or degradation of the hard drive which corrupts the stored classical bits. A real quantum computer will also have hardware errors. The quantum state of a qubit can change accidentally because of these hardware errors. Such errors may arise from the lack of full control of the interference between electromagnetic fields, variations in temperature, or energy dissipation. The accidental and incorrect change of a qubit state gives rise to the wrong answer which is called "noise".[4] As quantum computers only measure the state of a qubit, they cannot easily tell if the measurement is correct or incorrect. When we humans interpret these results, noise can cause confusion as to which answer is actually correct. Minimizing noise error is the greatest obstacle to building quantum computers.[5] For example, noise will cause the histogram in Fig. 6.2 to not

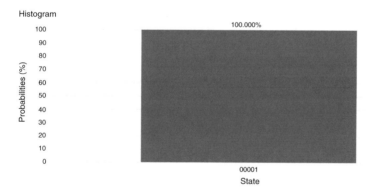

Fig. 6.2 Histogram showing that the qubit is measured in the $|1\rangle$ state with a probability of 1. Reprint Courtesy of International Business Machines Corporation. ©International Business Machines Corporation

---

[4]Background noise is an event that causes unwanted or incorrect affects on a signal.

[5]Noise can also occur in classical computers. Here, it can be because a wire in the computer which holds the 0- or 1-bit breaks and gives the wrong bit value. However, since classical computation has no probability associated with it, a single classical computation can be rerun twice and should give the exact same result. In practice, your computer reruns the same code many times to spot if there has been any errors and chooses the result which occurs most frequently. In this way you do not notice the hardware noise as easily.

have the perfect 100% outcome. Instead, noise will cause the qubit to be in the $|0\rangle$ state incorrectly some of the time, and the measurement histogram will incorrectly be $x\%$ in the $|0\rangle$ state and $(100 - x)\%$ in the $|1\rangle$ state. If the noise is large, then $x = 50\%$ and the measurement will be completely random. It should be understood that noise is an effect that occurs in both classical and quantum computers but because quantum computing technology is in its infancy, the noise is not as well under control.

6.3   ● Hadamard Gate

The Hadamard gate is very important in quantum computing. If the qubit starts in a definite $|0\rangle$ or $|1\rangle$ state, the Hadamard gate puts each into a superposition of $|0\rangle$ and $|1\rangle$ states. In Fig. 6.3, we apply a Hadamard gate to the $|0\rangle$ state qubit on the IBM Q simulator and measure the output.

The result of running the circuit 100 times is a histogram shown in Fig. 6.4. Note that each run is independent: before each measurement, the qubit has to be reset to the $|0\rangle$ state and passed through the gate, and then the measurement happens. We repeat this process 1024 times. Each bin in the histogram shows the frequency/probability of measuring $|0\rangle$ or $|1\rangle$. You can clearly see that applying the Hadamard gate to a single qubit creates a superposition state of both $|0\rangle$ and $|1\rangle$. The probabilities are not exactly 50/50 because of statistical error. The more data you collect, the closer the result converges to 50/50. This is similar to flipping a coin and counting the number of heads or tails; the greater the number of flips, the more likely you are to observe 50/50 probability of seeing heads/tails.

Recall that measurement collapses the superposition. Only one classical state can be observed, and all of the other quantum information is lost. Measurement collapse is the reason why a qubit's state cannot be duplicated, known as the no-cloning theorem of quantum computing. Once a superposition state is measured, it fundamentally changes into one of the basis states, and hence cannot be duplicated. Still, it is not known how or whether measurement collapse happens.[6]

Fig. 6.3 Applying a Hadamard gate and measuring on the IBM Q machine

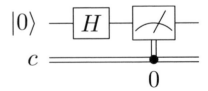

[6]https://en.wikipedia.org/wiki/Measurement_problem.

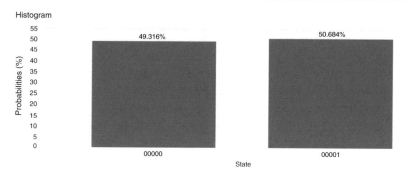

Fig. 6.4 Measurement histogram after running the Hadamard gate circuit in Fig. 6.3 1024 times. Reprint courtesy of International Business Machines Corporation, ©International Business Machines Corporation

$$|0\rangle -\boxed{H}-\boxed{H}- |0\rangle \quad |1\rangle -\boxed{H}-\boxed{H}- |1\rangle$$

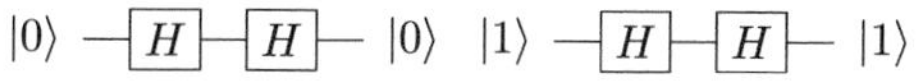

Fig. 6.5 Applying two Hadamard gates to the $|0\rangle$ state or $|1\rangle$ state

Question 1 Create a qubit in the $|1\rangle$ state and pass it through a Hadamard gate. From the measurement histogram, can you tell whether the qubit started in a $|0\rangle$ or $|1\rangle$ initial state?

The measurement histogram should look identical whether $|0\rangle$ or $|1\rangle$ was the initial state. Then how can we tell what the initial state was after a Hadamard operation? In the beam splitter, we determined where the photon came from by adding a second beam splitter to create interference. The way to measure and distinguish between them is to add a second Hadamard gate.

Question 2 Build a circuit that applies two Hadamard gates to a qubit in the $|0\rangle$ initial state as shown in Fig. 6.5. What is the output? Repeat this experiment for the $|1\rangle$ initial state.

6.4 ◆ Mathematics of the Hadamard Gate

The Hadamard gate has the following matrix representation:

$$H = \frac{1}{\sqrt{2}} \begin{pmatrix} 1 & 1 \\ 1 & -1 \end{pmatrix}. \tag{6.3}$$

Using matrix multiplication we can show that application of the Hadamard gate to an $|0\rangle$ initial state puts the qubit into the $(1/\sqrt{2})(|0\rangle + |1\rangle)$ state, also called the $|+\rangle$ state:

$$|0\rangle \boxed{H} \frac{1}{\sqrt{2}}\left(|0\rangle + |1\rangle\right). \tag{6.4}$$

If the initial state is $|1\rangle$, the Hadamard gate will create the superposition $(1/\sqrt{2})(|0\rangle - |1\rangle)$ state, called the $|-\rangle$:

$$|1\rangle \boxed{H} \frac{1}{\sqrt{2}}\left(|0\rangle - |1\rangle\right). \tag{6.5}$$

In the Stern–Gerlach experiment, you learned that the $|0\rangle$ and $|1\rangle$ states make up the z-basis and are associated with spin up and spin down. The $|+\rangle$ and $|-\rangle$ states comprise the x-basis and are associated with spin right and spin left. While the Stern–Gerlach could be rotated to measure at any angle, a quantum computer is physically built to only measure in the z-basis. Therefore, the spin right $1/\sqrt{2}(|0\rangle + |1\rangle)$ and spin left $1/\sqrt{2}(|0\rangle - |1\rangle)$ look the same when measured by a quantum computer. However, the two states have hidden information that can be recovered by using a second Hadamard gate to change back into the z-basis.

6.4.1 Examples

1. A spin right $1/\sqrt{2}(|0\rangle + |1\rangle)$ is sent through a Hadamard gate, creating a superposition of $|+\rangle$ and $|-\rangle$ given by $1/\sqrt{2}(|+\rangle + |-\rangle)$. By performing a basis change, show that this is equivalent to producing a $|0\rangle$ state.

$$\frac{1}{\sqrt{2}}\left(|+\rangle + |-\rangle\right) = \frac{1}{\sqrt{2}}\left(\frac{1}{\sqrt{2}}|0\rangle + \frac{1}{\sqrt{2}}|1\rangle\right) + \frac{1}{\sqrt{2}}\left(\frac{1}{\sqrt{2}}|0\rangle - \frac{1}{\sqrt{2}}|1\rangle\right), \tag{6.6}$$

$$= \frac{1}{2}|0\rangle + \frac{1}{2}|1\rangle + \frac{1}{2}|0\rangle - \frac{1}{2}|1\rangle, \tag{6.7}$$

$$= |0\rangle. \tag{6.8}$$

2. Use matrix multiplication to show how applying the Hadamard gate twice to a $|0\rangle$ state qubit recovers its original state.

$$H|0\rangle = \frac{1}{\sqrt{2}}\begin{pmatrix} 1 & 1 \\ 1 & -1 \end{pmatrix}\begin{pmatrix} 1 \\ 0 \end{pmatrix} = \frac{1}{\sqrt{2}}\begin{pmatrix} 1 \\ 1 \end{pmatrix}, \tag{6.9}$$

$$HH|0\rangle = \frac{1}{2} \begin{pmatrix} 1 & 1 \\ 1 & -1 \end{pmatrix} \begin{pmatrix} 1 \\ 1 \end{pmatrix} = \begin{pmatrix} 1 \\ 0 \end{pmatrix}. \qquad (6.10)$$

In fact, all quantum gates are reversible as a consequence of the unitary matrix condition. Recall that the gates must be unitary so that the probabilities always add up to 1. Multiplying any unitary matrix by its conjugate transpose will return the identity matrix, i.e., reverses the gate to get the original state by $UU^\dagger = U^\dagger U = 1$. The Hadamard matrix is self-unitary, i.e., it is its own conjugate transpose, $U = U^\dagger$.

6.5   ■ Z Gate

The Z-gate matrix representation is

$$Z = \begin{pmatrix} 1 & 0 \\ 0 & -1 \end{pmatrix}. \qquad (6.11)$$

The Z gate leaves a $|0\rangle$ state unchanged but flips the sign of the $|1\rangle$ state to $-|1\rangle$ by

$$\alpha|0\rangle + \beta|1\rangle \; -\boxed{Z}- \; \alpha|0\rangle - \beta|1\rangle. \qquad (6.12)$$

This is equivalent to changing the qubit from a $|+\rangle$ state to a $|-\rangle$ state. The effects of the X, H, and Z gates are summarized in Fig. 6.6.

Fig. 6.6 The X, H, and Z gates change the qubit's state in the z- and x-basis and are related according to this diagram

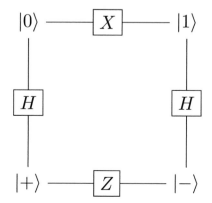

6.6   Big Ideas

1. Every interaction with a classical computer is caused by code instructing classical logic gates to operate on classical bits. Similarly, a calculation on a quantum computer is caused by coding quantum logic gates to act on qubits.
2. Common single-qubit gates include the X, Z, and H (Hadamard) gates.
3. Each quantum gate can be mathematically represented as a unitary matrix which acts on qubits.

6.7   Activities

● Exploring gates on the IBM Quantum Computer 10.4.

6.8   Check Your Understanding

1. ■ Use matrix multiplication to show how applying an X gate flips:
    (a) A qubit in the $|0\rangle$ state.
    (b) A qubit in the general $|\psi\rangle = \alpha|0\rangle + \beta|1\rangle$ state.
2. ● Explain the relationship between a beam splitter and a Hadamard gate.
3. ● A $|0\rangle$ qubit is passed through a Hadamard gate. We measure the qubit state as $|1\rangle$. Which of the following choices best describes the result if we perform a measurement on the qubit a second time without reinitializing?
    (A) $|0\rangle$
    (B) $|1\rangle$
    (C) 50% chance of $|0\rangle$ or $|1\rangle$
4. ● Assume a qubit represents a light bulb that can be measured as either ON or OFF.
    (a) The light bulb is originally ON. What gate would you use to turn it OFF?
    (b) The light bulb is originally ON and passes through a Hadamard gate. What do you measure as the output?
    (c) The light bulb is originally ON and passed through two Hadamard gates in series. What do you measure as the output?
5. ● Explain how the Hadamard gate is implemented in the Stern–Gerlach experiment.
6. ■ Explain the output of the Mach–Zehnder interferometer using what you learned about Hadamard gates.
7. ■ Use matrix multiplication to demonstrate
    (a) The Hadamard gate applied to a $|1\rangle$ state qubit turns it into a $|-\rangle$.
    (b) A second Hadamard gate turns it back into the $|1\rangle$ state.
    (c) The output after applying the Hadamard gate twice to a general state $|\psi\rangle = \alpha|0\rangle + \beta|1\rangle$.

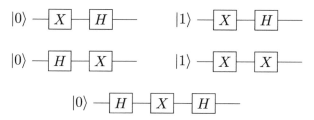

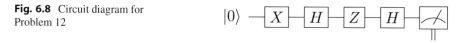

Fig. 6.7 Five quantum circuits for Problem 8

Fig. 6.8 Circuit diagram for Problem 12

8. ● Which of the quantum circuits in the Fig. 6.7 would NOT produce the histogram shown in Fig. 6.4?

9. ◆ Use matrix multiplication to show how applying the Z gate to $|+\rangle$ changes it to $|-\rangle$.

10. ■ Using only Hadamard and Z gates, design a quantum circuit that outputs the same result as an X gate.

11. ◆ Using the IBM Q simulator, apply the Z gate to a qubit in the following initial states and interpret the measurement histogram.
    (a) $|0\rangle$
    (b) $|1\rangle$ (Hint: You need to first flip the $|0\rangle$ state using the X gate.)
    (c) $|+\rangle$ (Hint: You need to first create the $|+\rangle$ state using the H gate.)
    (d) $|-\rangle$ (Hint: You need to first create the $|-\rangle$ state using the X and H gates.)

12. ■ What is the expected measurement histogram produced by the circuit in Fig. 6.8?

13. ◆ Show that the Hadamard gate is unitary and therefore reversible.

Entanglement

7

So far, we have discussed the manipulation and measurement of a single qubit. However, **quantum entanglement** is a physical phenomenon that occurs when multiple qubits are correlated with each other. Entanglement can have strange and useful consequences that could make quantum computers faster than classical computers. Qubits can be "entangled," providing hidden quantum information that does not exist in the classical world. It is this entanglement that is one of the main advantages of the quantum world!

7.1 ● Entanglement Fundamentals

To provide one example of the strange behavior of entanglement, suppose we have two fair coins. Classically, if you flipped two fair coins many times, you would measure the outcomes HH, HT, TH, or TT, each occurring with a 25% probability. However, by quantum entangling these two fair coins, it is possible to create a state $(1/\sqrt{2})(|HH\rangle + |TT\rangle)$ as illustrated in Fig. 7.1. Many other types of entangled states are possible, but this is one famous example called a Bell state. If you flipped this "entangled" pair of coins, they are entangled in such a way that only two measurement outcomes are possible: (1) both coins land on heads; or (2) both coins land on tails; each outcome occurring with 50% probability. You would never see HT or TH!

Furthermore, if the two entangled coins are separated by thousands of miles, one coin can be flipped and measured. In this case, if the measured coin produced the outcome heads, then we automatically know that the other coin must also land on heads. If the measured coin produced the outcome tails, then we automatically know that the other coin must also land of tails! If this isn't strange enough, this appears to suggest that the two coins can transmit information instantaneously, and possibly even faster than the speed of light (the fastest speed in the Universe), as shown in Fig. 7.2. If the two coins are flipped at the exact same time, somehow the two coins

© The Author(s) 2021
C. Hughes et al., *Quantum Computing for the Quantum Curious*,
https://doi.org/10.1007/978-3-030-61601-4_7

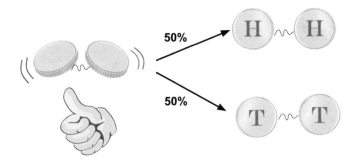

Fig. 7.1 Two coins that are entangled in such a way that they either both land on HH or both land on TT.

Fig. 7.2 Two coins are separated with no means of communication between each other. Classically, the flip of the second coin would be unrelated to the first flip. However, entangled coins would still produce correlated results.

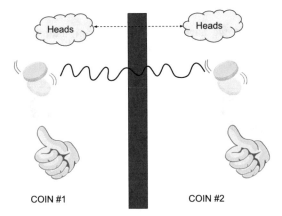

know to land on the same side as the other even though there can be no classical communication between them.

How does the other coin instantaneously "know" what was measured on the other? Is information somehow being transmitted faster than the speed of light? Einstein called this behavior a "spooky action at a distance."[1] It has since been shown that no information is being transmitted from one place to the other, and so no information is being transmitted faster than the speed of light. Rather, the particles share non-classical information at the time of entanglement, which is then observed in the measurement process. The correlation between entangled qubits is the key that allows quantum computers to perform certain computations much faster than classical computers.

---

[1] "Bounding the speed of spooky action at a distance." *Physical Review Letters*. 110: 260407. 2013. arXiv:1303.0614.

7.2 ● Hidden Variable Theory

It is tempting to think that there may be some classical explanation for entanglement. Did the entanglement change the fair coins by adding extra mass to the heads side or the tails side, thereby making them unfair? To provide a more realistic example in a classical system, consider a particle that decays into two lighter particles. The momenta of these three particles are related by the conservation of momentum: $\vec{p}_i = \vec{p}_{f1} + \vec{p}_{f2}$. Given a known total initial momentum, then by measuring the momentum of one of the final state particles, we can determine the momentum of the other final state particle. In summary, by measuring one particle's momentum, we know the other. Momentum is the hidden classical variable that is encoded when the two particles are created. This is shown in Fig. 7.3. Naturally, the question arises: is there a conceptually similar hidden variable in the quantum mechanical situation?

However, Bell's theorem[2] demonstrated that the correlation between entangled quantum particles is more than what is possible classically, disproving the idea of a hidden variable. All other potential loopholes have been resolved as of 2016.[3] As such, entanglement is a purely quantum phenomenon with no classical explanation.

7.3 ● Multi-Qubit States

Given multiple qubits, the total state of the system can be written together in a single ket. For example, if coin #1 is heads and coin #2 is tails, the two-coin state is expressed as $|HT\rangle$. In general, a system of two qubits can be in a superposition of four classical states, and written as

$$|\psi\rangle = \alpha_{00}|00\rangle + \alpha_{01}|01\rangle + \alpha_{10}|10\rangle + \alpha_{11}|11\rangle.$$

Fig. 7.3 When a particle decays into two smaller particles, the decay products are "classically entangled" according to the conservation of momentum.

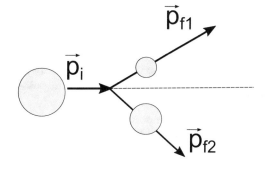

---

[2]https://brilliant.org/wiki/bells-theorem/.

[3]The BIG Bell Test Collaboration (9 May 2018). "Challenging local realism with human choices." *Nature*. 557: 212–216. doi:10.1038/s41586-018-0085-3.

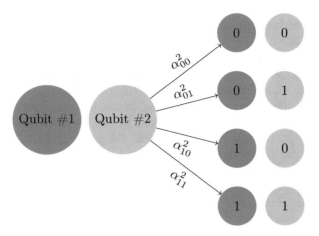

Fig. 7.4 A two-qubit system can collapse into one of four states with probability α_{ij}^2.

As we saw for the single qubit states, the coefficients α_{ij} are called the amplitudes and are generally complex numbers. Measuring the two qubits will collapse the system into one of the four basis states with probability given by α_{ij}^2. This is shown in Fig. 7.4.

7.3.1   Example

A system of two qubits is in a superposition state given by $|\psi\rangle = \frac{1}{\sqrt{2}}|00\rangle + \frac{1}{2}|10\rangle - \frac{1}{2}|11\rangle$.

(a) What is the probability of measuring both qubits as 1?

Prob $(|11\rangle) = \left(-\frac{1}{2}\right)^2 = \frac{1}{4}$.

(b) If we only measure the first qubit and get a value of 1, what is the new state of the system?

Since $|00\rangle$ is the only basis state of $|\psi\rangle$ that doesn't have a 1 in the first qubit, we eliminate the state $|00\rangle$ from the possibilities. This results in $|\psi'\rangle = \frac{1}{2}|10\rangle - \frac{1}{2}|11\rangle$.

Finally, we re-normalize the state so that the probabilities add up to 1. Therefore, the new state is $|\psi'\rangle = \frac{1}{\sqrt{2}}|10\rangle - \frac{1}{\sqrt{2}}|11\rangle$.

7.4   ⬤ Non-Entangled Systems

It is possible to have a system of particles that are not entangled with each other. In this case, changing one particle will not cause any change in the other particle. For example, in a classical system, flipping two coins and measuring one coin as heads does not tell you any information about whether or not the other coin will land on heads or tails. These events are said to be independent. If you wanted to calculate the probability of $|HT\rangle$, you would simply multiply the probability of getting H on coin #1 by the probability of getting T on coin #2. This is given by

$$\text{Prob}\,(|HT\rangle) = \left(\frac{1}{2}\right)\left(\frac{1}{2}\right) = \frac{1}{4}.$$

Non-entangled states are also called product states or separable states because they can be factored into a product of single-qubit states.[4] The two single-qubit probabilities multiply to produce the two-qubit probabilities.

7.4.1 Example

One qubit is in a $\alpha_0|0\rangle + \alpha_1|1\rangle$ state, while another is in a $\beta_0|0\rangle + \beta_1|1\rangle$ state. What is the state of the non-interacting two-qubit system?

$$(\alpha_0|0\rangle + \alpha_1|1\rangle)\,(\beta_0|0\rangle + \beta_1|1\rangle) = \alpha_0\beta_0|00\rangle + \alpha_0\beta_1|01\rangle + \alpha_1\beta_0|10\rangle + \alpha_1\beta_1|11\rangle.$$

7.5 ■ Entangled Systems

Intuitively, any interaction between two or more qubits will cause the qubits to share some information between each other. This sharing of information from interactions causes emergent phenomena that we call entanglement. Mathematically, a multi-qubit state is entangled only if it cannot be expressed as a product state. However, determining if a general multi-qubit state can be expressed as a product state can be difficult. Instead, an easier test to determine if a system is entangled or not is to check if measuring the value of one qubit changes the probability distribution of the second qubit. We will use this test extensively.[5] If this test is true, then the system is definitely entangled. However, if the probability distribution of the second qubit does not change in this test, then the system could still be entangled. In this case, the entanglement arises from the sharing of hidden information in the signs (or complex components) of the probability amplitudes of a general state. While the probabilities may not change (due to the amplitudes being squared), the relative

---

[4]More recently, it has been shown that there can exist quantum correlations in separable states that are not due to entanglement. These are called quantum discord: https://en.wikipedia.org/wiki/Quantum_discord.

[5]There are ways to test for entanglement without the need to factorise a multi-qubit state into single qubit states. Namely, whether the trace of the density matrix for the subsystem squared is equal to 1. However, the mathematical necessities for this test are outside the scope of this course.

signs have important consequences for constructive or destructive interference if more gates are applied to qubits. This concept will be explored in question 5 e).

7.5.1 Example

Is $|\psi\rangle = \frac{1}{\sqrt{2}}|00\rangle + \frac{1}{\sqrt{2}}|11\rangle$ an entangled state?

Yes! To see this, examine qubit #2. The probabilities for measuring qubit #2 in the $|0\rangle$ or $|1\rangle$ states are originally 50/50 respectively. However, if we measured qubit #1, then we know what the outcome of measuring qubit #2 will be with 100% certainty. The same argument holds if qubit #2 is measured first. As such, measuring one of the qubits affects the probability of measuring the other qubit in a certain state, and so they are entangled. Mathematically, an entangled state is a special multi-qubit superposition state that cannot be factored into a product of the individual qubits.

7.5.2 Example

Show that $|\psi\rangle = \frac{1}{\sqrt{2}}|00\rangle + \frac{1}{\sqrt{2}}|11\rangle$ cannot be written as a product of two single qubits.

Assume that the state can be written as the product of two states.

$$\frac{1}{\sqrt{2}}|00\rangle + \frac{1}{\sqrt{2}}|11\rangle \stackrel{?}{=} (\alpha_0|0\rangle + \alpha_1|1\rangle)(\beta_0|0\rangle + \beta_1|1\rangle), \tag{7.1}$$

$$\stackrel{?}{=} \alpha_0\beta_0|00\rangle + \alpha_0\beta_1|01\rangle + \alpha_1\beta_0|10\rangle + \alpha_1\beta_1|11\rangle. \tag{7.2}$$

Comparing the amplitudes on the left vs. the right, the α_i's and β_j's must satisfy:

$$\alpha_0\beta_0 = \frac{1}{\sqrt{2}}, \quad \alpha_0\beta_1 = 0, \quad \alpha_1\beta_0 = 0, \quad \alpha_1\beta_1 = \frac{1}{\sqrt{2}}. \tag{7.3}$$

However, this is not possible. For example, take $\alpha_0\beta_1 = 0$. This means that either $\alpha_0 = 0$ or $\beta_1 = 0$. If $\alpha_0 = 0$, then $\alpha_0\beta_0 = 0$, but $\alpha_0\beta_0 = \frac{1}{\sqrt{2}}$ in the above equation. A similar contradiction occurs with $\beta_1 = 0$. So the initial assumption must be incorrect and this entangled state cannot be written as the product of two separate states.

7.6   ■ Entangling Particles

As there are many different ways of building a quantum computer, there are many different ways of physically entangling particles. One method called "spontaneous parametric down-conversion" shines a laser at a special nonlinear crystal. The

Fig. 7.5 A nonlinear crystal creates two photons with entangled polarizations.

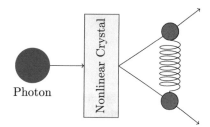

Table 7.1 The truth table for the CNOT gate

Before		After	
Control bit	Target bit	Control bit	Target bit
$|0\rangle$	$|0\rangle$	$|0\rangle$	$|0\rangle$
$|0\rangle$	$|1\rangle$	$|0\rangle$	$|1\rangle$
$|1\rangle$	$|0\rangle$	$|1\rangle$	$|1\rangle$
$|1\rangle$	$|1\rangle$	$|1\rangle$	$|0\rangle$

crystal splits the incoming photon into two photons with correlated polarizations. For example, one could produce a pair of photons that always have perpendicular polarizations (see Fig. 7.5). Just as the engineering aspect of building a quantum computer is outside the scope of this course, so is the technological aspect of how qubits are physically entangled. We will focus more on how entanglement is represented in a quantum computer and the uses of this.

7.7 ■ CNOT Gate

You have already learned about the X, Hadamard, and Z gates. These act on a single qubit. There are also quantum gates that perform a logic operation on multiple qubits. The most important multi-qubit gate is the controlled NOT (CNOT) gate. The CNOT is used to entangle two qubits together and is essential in quantum computing/algorithms. The CNOT takes in two qubits, a control qubit and a target qubit, and outputs two qubits. The control qubit stays the same, while the target obeys the following rule.

- If the control qubit is $|0\rangle$, then leave the target qubit alone.
- If the control qubit is $|1\rangle$, then on the target qubit flip $|0\rangle \rightarrow |1\rangle$ and $|1\rangle \rightarrow |0\rangle$.

The truth table for the CNOT gate is shown in Table 7.1.[6] From this one can deduce the matrix form of the CNOT gate as

---

[6]https://en.wikipedia.org/wiki/Controlled_NOT_gate.

$$\mathrm{CNOT} = \begin{pmatrix} 1\,0\,0\,0 \\ 0\,1\,0\,0 \\ 0\,0\,0\,1 \\ 0\,0\,1\,0 \end{pmatrix}. \tag{7.4}$$

Figure 7.6 is the circuit for the CNOT gate. Plugging in the "Before" values from Table 7.1 into this circuit will produce the "After" values.

7.8 ■ Notation Convention

When converting between bra-ket notation and circuit notation, there are two different conventions. Since we will be using the IBM quantum computer, we will adopt the IBM convention. This is shown in Fig. 7.7. In the IBM notation, the topmost qubit in the circuit corresponds to the rightmost qubit in the bra-ket notation ($|\ldots q\rangle$). IBM shorthand is top-down in circuit notation, which corresponds to right-left in bra-ket notation.

The other convention, which we will **not** use going forward but provide in case it is seen in other resources, is shown in Fig. 7.7. Here, the topmost qubit corresponds to the leftmost qubit in bra-ket notation ($|q\ldots\rangle$). Top-down in circuit notation corresponds to left-right in bra-ket notation. We will not use this going forward.

7.9 Examples

1. Figure 7.8 shows the quantum circuit sending $|01\rangle$ through a CNOT gate. What is the output?

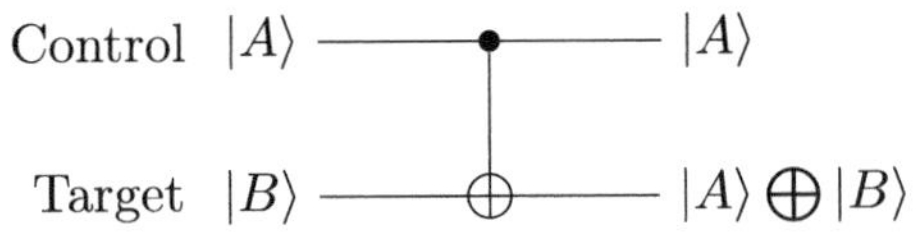

Fig. 7.6 The CNOT gate performs an X gate on the target qubit if the control qubit is $|1\rangle$.

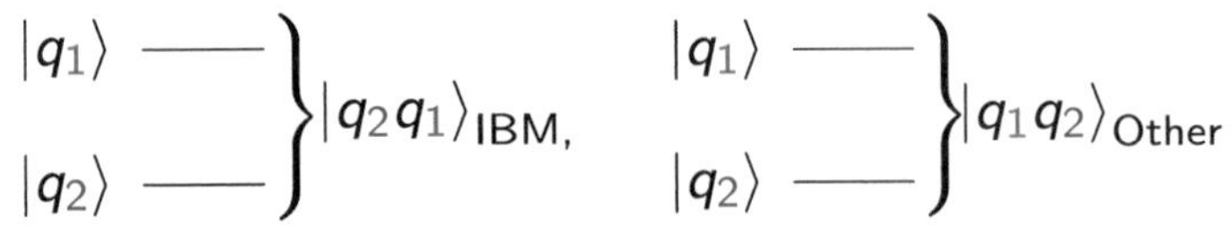

Control $|A\rangle$ ——————•—————— $|A\rangle$

Target $|B\rangle$ ——————⊕—————— $|A\rangle \oplus |B\rangle$

$|q_1\rangle$ ———
$|q_2\rangle$ ——— $\Big\} |q_2 q_1\rangle_{\text{IBM}}$,

$|q_1\rangle$ ———
$|q_2\rangle$ ——— $\Big\} |q_1 q_2\rangle_{\text{Other}}$

Fig. 7.7 The two conventions for mapping circuit notation to the bra-ket notation, IBM (left) and Other (right). Note in this book we adopt the IBM convention as we run code on the IBM quantum computers.

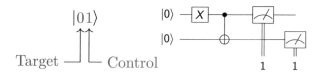

Target — Control

Fig. 7.8 The quantum circuit that sends a multi-qubit in the $|01\rangle$ state through a CNOT gate.

Fig. 7.9 The quantum circuit that sends a control qubit in a superposition state through a CNOT gate.

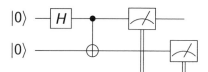

The figure shows that, in IBM notation, the control qubit is on top and the target is on the bottom. Since the control is in the $|1\rangle$ state, the target qubit is flipped to $|1\rangle$. So measurement will always result in $|11\rangle$.

2. Examine Fig. 7.9. The control qubit is in a superposition of $|0\rangle$ and $|1\rangle$. What is the effect of a CNOT gate?

Before the CNOT operation, in ket notation, the control qubit is in the $\frac{1}{\sqrt{2}}|0\rangle + \frac{1}{\sqrt{2}}|1\rangle$ state, while the target qubit is in the $|0\rangle$ state. The two-qubit input state is therefore $\frac{1}{\sqrt{2}}|00\rangle + \frac{1}{\sqrt{2}}|01\rangle$. Applying the rules for the CNOT, the first state $|00\rangle$ does not change as the control qubit is $|0\rangle$. However, for the second state $|01\rangle$, the control qubit is $|1\rangle$ and so the target qubit is flipped from $|0\rangle$ to $|1\rangle$. The result of the CNOT gate is the state $\frac{1}{\sqrt{2}}|00\rangle + \frac{1}{\sqrt{2}}|11\rangle$. The histogram from measuring this state is shown in Fig. 7.10. This is a special state called the Bell state.

The two qubits are entangled after the CNOT! As illustrated in the previous example, this state cannot be written as the product of two separate qubits. As with the single-qubit gates, the CNOT gate operates on ALL states in the superposition, e.g., the CNOT gate acts on the four basis states of a two qubit system simultaneously. Quantum algorithms leverage this parallelism to ensure speed improvements over classical computers. In addition, as with all quantum gates, the CNOT is reversible, meaning the operation can be undone (which can be used to figure out the original qubit states).

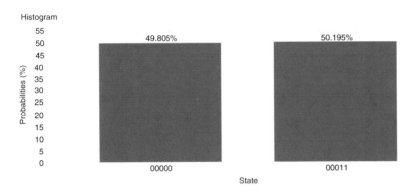

Fig. 7.10 The measurement histogram produced by running the circuit in Fig. 7.9. Reprint courtesy of International Business Machines Corporation, ©International Business Machines Corporation.

7.10   Big Ideas

1. Entanglement is the sharing of non-classical information between two or more quantum states. This is caused by quantum states or qubits interacting with each other.
2. Entanglement is needed to make quantum computers perform calculations which classical computers cannot.
3. Two-qubit gates act on two different qubits simultaneously and creates entanglement. The controlled NOT (CNOT) gate is frequently used for this purpose.[7]

7.11   Activities

◼ Correlation in Entangled States Lab in Worksheet 10.1
● Schrödinger's Worm Using Five Qubits in Worksheet 10.4
● For those interested in hands-on experiments, see QuTools[8]

7.12   Check Your Understanding

1. ● For each of the questions below, assume that two-qubits start in the state

$$|\psi\rangle = \frac{1}{\sqrt{2}}|00\rangle + \frac{1}{2}|10\rangle - \frac{1}{2}|11\rangle. \tag{7.5}$$

---

[7]In fact, three single qubit gates (the Hadamard, phase, and $\pi/8$ phase-rotation) in combination with the CNOT form a universal set of gates, i.e., all other gates can be made up from them.

[8]https://www.qutools.com/quantum-physics-education-science-kits/.

Fig. 7.11 CNOT gate for
Problem 6.

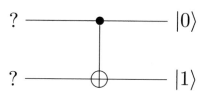

(a) What is the probability of measuring both qubits as 0?
(b) What is the probability of measuring the first qubit as 1?
(c) What is the probability of measuring the second qubit as 0?
(d) What is the new state of the system after measuring the first qubit as 0?
(e) What is the new state of the system after measuring the first qubit as 1?

2. ⬤ Two fair coins are flipped. What is the state of the two-coin system while the coins are in the air?

3. ◼ Is $\frac{1}{\sqrt{2}}|00\rangle + \frac{1}{\sqrt{2}}|01\rangle$ an entangled state? If so, show that it cannot be written as a product. If not, what is the individual state of the two qubits?

4. ◆ Are the following two-qubit states entangled?
   (a) $\frac{1}{\sqrt{2}}|01\rangle + \frac{1}{\sqrt{2}}|10\rangle$
   (b) $\frac{1}{\sqrt{2}}|01\rangle - \frac{1}{\sqrt{2}}|10\rangle$
   (c) $\frac{\sqrt{3}}{2}|00\rangle + \frac{1}{2}|11\rangle$
   (d) $\frac{1}{\sqrt{2}}|10\rangle + \frac{1}{\sqrt{2}}|11\rangle$
   (e) $\frac{1}{2}|00\rangle + \frac{1}{2}|01\rangle + \frac{1}{2}|10\rangle - \frac{1}{2}|11\rangle$
   (f) $\frac{1}{\sqrt{2}}|00\rangle + \frac{1}{2}|10\rangle - \frac{1}{2}|11\rangle$

5. ◼ Two qubits are passed through a CNOT. In IBM notation, the qubit on the right is the control qubit. What is the output for the following initial states?
   (a) $|00\rangle$
   (b) $|01\rangle$
   (c) $|11\rangle$
   (d) $\frac{1}{\sqrt{2}}|01\rangle + \frac{1}{\sqrt{2}}|10\rangle$
   (e) $\frac{1}{\sqrt{2}}|00\rangle + \frac{1}{2}|10\rangle - \frac{1}{2}|11\rangle$

6. ◼ The output of a CNOT gate is shown in Fig. 7.11. What were the inputs?

7. ◼ Can you predict the state produced by these quantum circuits? Try them out on the IBM quantum computer.

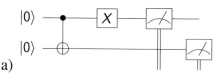
a)

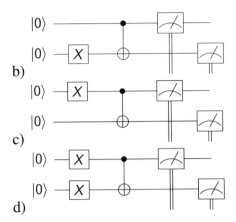

8. ■ Can you predict which states will be produced by these quantum circuits?

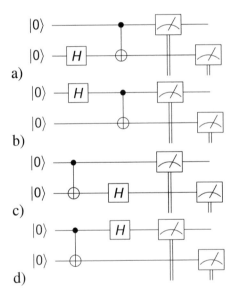

9. ◆ Can you predict the state produced by these quantum circuits? Note: to put the circuit in a) on the IBM quantum computer, you need to use the code (rather than the click-and-drag interface) to put the second CNOT control on the bottom qubit.

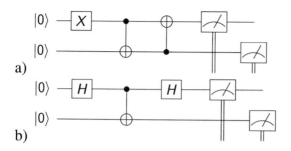

10. ■ Use the IBM Q[9] simulator to create the entangled state $\frac{1}{\sqrt{2}}|01\rangle + \frac{1}{\sqrt{2}}|10\rangle$.

11. ■ Suppose Alice has one half of an entangled pair and Bob has the other half. When Alice makes a measurement on her qubit, Bob's qubit instantaneously changes its state. Can Alice and Bob use entanglement to transmit information faster than the speed of light? Why or why not?

---

[9] https://quantum-computing.ibm.com/

Quantum Teleportation

8

One interesting application of entanglement is **quantum teleportation**, which is a technique for transferring an *unknown* quantum state from one place to another. In science fiction, teleportation generally involves a machine scanning a person and another machine reassembling the person on the other end. The original body disintegrates and no longer exists. Similarly, quantum teleportation works by "scanning" the original qubit, sending a recipe, and reconstructing the qubit elsewhere. The original qubit is not physically destroyed in the science fiction sense, but it is no longer in the same state. Otherwise, the previously mentioned **no-cloning** theorem—which states that a qubit cannot be exactly copied onto another qubit— would be violated.[1] As we will see, the "scanning" part poses a problem which can only be solved by leveraging quantum entanglement.

8.1 ■ Scanning a Qubit

Question 1 Create a qubit in the $|1\rangle$ state and pass it through a Hadamard gate. From the measurement histogram, can you tell whether the qubit started as a $|0\rangle$ or $|1\rangle$ initial state?

The measurement histogram should look identical if either of the $|0\rangle$ or $|1\rangle$ states is used initially. Then how can we tell what the initial state was after performing a Hadamard operation? In the beam splitter, we determined where the photon came from by adding a second beam splitter to create interference. The way to measure and distinguish between them is to add a second Hadamard gate. As we have seen in Sect. 2.2, all gates must be unitary to conserve probabilities. The unitary condition ensures that all gates are reversible: we can undo the action of any gate by applying

---

[1] The no-cloning theorem poses a big problem for correcting errors that happen on quantum computers: https://en.wikipedia.org/wiki/Quantum_error_correction.

© The Author(s) 2021
C. Hughes et al., *Quantum Computing for the Quantum Curious*,
https://doi.org/10.1007/978-3-030-61601-4_8

its conjugate transpose. This is easily seen in matrix form as unitary matrices as defined as $UU^{\dagger} = \mathbb{1}$. As the Hadamard gate is its own conjugate transpose, applying a second Hadamard gate is equivalent to undoing the first. This is how the original state is recovered.

Question 2 If a qubit is in the unknown state $a|0\rangle + b|1\rangle$, what is the result of a single measurement?

(A) 0
(B) 1
(C) 0 with probability a^2 and 1 with probability b^2
(D) A number between 0 and 1

Question 3 What is the result of a second measurement after the first from Question 2?

(A) 0 if the first measurement is 0 or 1 if the first measurement is 1
(B) 0 if the first measurement is 1 or 1 if the first measurement is 0
(C) 0 with probability a^2 and 1 with probability b^2
(D) A number between 0 and 1

Given a single qubit, it is not possible to determine how much of a superposition it is in if you only have this single qubit, i.e., you cannot determine the coefficients of $|0\rangle$ and $|1\rangle$ in a general state from one measurement! Note that if the state is known (from measuring many independent qubits that have been prepared identically), then you can just directly send the recipe to prepare this qubit. It is only when the state is unknown and when there is only one qubit that we have to think harder about how to efficiently "scan" the particle.

8.2   ■ Teleportation Protocol

The way to get around the problem of not being able to measure the qubit (and avoid collapsing the unknown state onto a basis state) is to "scan" the qubit indirectly with the help of entangled particles. This comic[2] illustrates the basic idea. The protocol is as follows:

1. Alice and Bob meet up and make a qubit each (which we will call qubit #2 and #3). At this point, the two qubits are completely independent and we can think of the qubits as two different balls that do not contain any information about the other. Then, Alice and Bob decide to entangle their qubits by causing an interaction between the qubits, for example by applying a CNOT gate. Think

---

[2]https://www.jpl.nasa.gov/news/news.php?feature=4384.

of entanglement as Alice writing some information on Bob's ball that only she knows how to read, and Bob writing information on Alice's ball that only he knows how to read. For Bob to read Alice's information on his ball, Alice needs to send him a (classical) message describing how to understand it, and vice-versa. They do not tell each other how to read the information yet. One possible entangled state (called the Bell-state) that they decide to create is

$$\frac{1}{\sqrt{2}}|00\rangle + \frac{1}{\sqrt{2}}|11\rangle. \tag{8.1}$$

Alice takes her qubit and walks away, and Bob takes his and walks in a different direction as shown in Fig. 8.1.

2. Now Alice obtains a third qubit in an unknown state (qubit #1) that she wants to transfer to Bob. She can only communicate with him classically by email or phone, and it would take too long to physically bring the qubit to Bob. The current situation is shown in Fig. 8.2.

3. Alice interacts her two qubits using a CNOT gate (qubits #1 and #2) and measures the qubit she originally had (qubit #2). She then sends the unknown qubit to be teleported (qubit #1) through a Hadamard gate and afterwards measures the output. Recall that the Hadamard gate is used to create a superposition of states. The current situation is shown in Fig. 8.3.

   Because Alice's original qubit (qubit #2) was entangled with Bob's, the CNOT interaction with qubit #1 immediately changes the state of Bob's qubit.

   When understanding quantum teleportation, it may be more insightful to see the mathematical description of this three-qubit protocol. Qubit #1 is the qubit to be teleported, and qubit #2 and #3 are the entangled pair shared by Alice and Bob. In ket notation, the three-qubit state is written in the order $|\,\#1\;\#2\;\#3\,\rangle$. In addition, the three-qubit state can be written in ket notation in different ways as

Fig. 8.1 Alice and Bob's qubits are entangled

Fig. 8.2 Alice has a qubit (#1) in an unknown state she wants to transfer to Bob

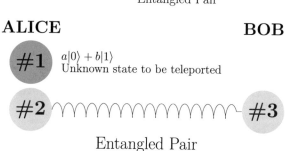

ALICE

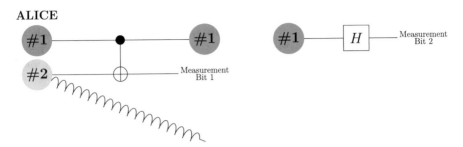

Fig. 8.3 Alice passes her two qubits through a CNOT gate

long as the order of the qubits is kept unchanged. For example, if qubit #1 = $|0\rangle$, qubit #2 = $|1\rangle$, and qubit #3 = $|1\rangle$, they can be written as $|0\rangle|1\rangle|1\rangle = |011\rangle = |0\rangle|11\rangle = |01\rangle|1\rangle$. Sometimes it is easier to split up the multi-qubit state like this to explicitly show if a gate is acting on a single qubit.

Now, from steps (1) and (2) we have Alice and Bob's qubits in the Bell state $\frac{1}{\sqrt{2}}(|00\rangle + |11\rangle)$. Qubit #1 is in an unknown state $a|0\rangle + b|1\rangle$. At the start of step (3), the three qubits need to be written together in ket notation by multiplying the qubit to be teleported by the entangled Bell state. The product is:

$$[a|0\rangle + b|1\rangle]\left(\frac{1}{\sqrt{2}}|00\rangle + \frac{1}{\sqrt{2}}|11\rangle\right) = \frac{1}{\sqrt{2}}\left(a|000\rangle + a|011\rangle + b|100\rangle + b|111\rangle\right).$$
(8.2)

Next, we apply a CNOT gate using the first qubit in Eq. (8.2) as the control and the second as the target. Recall that the target (qubit #2) changes state only if the control is $|1\rangle$. After applying the CNOT gate the three-qubit state is

$$\frac{1}{\sqrt{2}}\left(a|000\rangle + a|011\rangle + b|110\rangle + b|101\rangle\right).$$
(8.3)

After this, we apply a Hadamard gate to qubit #1 in Eq. (8.3). Recall that the Hadamard gate changes the state $|0\rangle \rightarrow (1/\sqrt{2})(|0\rangle + |1\rangle)$, and $|1\rangle \rightarrow (1/\sqrt{2})(|0\rangle - |1\rangle)$. The three-qubit state is

$$\frac{1}{2}\left(a\left(|0\rangle + |1\rangle\right)|00\rangle + a\left(|0\rangle + |1\rangle\right)|11\rangle + b\left(|0\rangle - |1\rangle\right)|10\rangle + b\left(|0\rangle - |1\rangle\right)|01\rangle\right).$$
(8.4)

Next, distribute the product of qubits throughout Eq. (8.4) to find

$$\frac{1}{2}\left(a|000\rangle + a|100\rangle + a|011\rangle + a|111\rangle + b|010\rangle - b|110\rangle + b|001\rangle - b|101\rangle\right).$$
(8.5)

Finally, combine like-terms of Eq. (8.5) based on the first two qubits to get

$$\frac{1}{2}\Big(|00\rangle(a|0\rangle+b|1\rangle)+|10\rangle(a|0\rangle-b|1\rangle)+|01\rangle(a|1\rangle+b|0\rangle)+|11\rangle(a|1\rangle-b|0\rangle)\Big)$$
(8.6)

Remember that qubit #1 and qubit #2 are the ones that belong to Alice. We see that the state of Bob's qubit #3 has changed by applying the CNOT and Hadamard gates to Alice's qubits. As can be seen in Eq. (8.6), Bob's qubit is currently in one of four possible superposition states. This is shown in Fig. 8.4.

The four possible superposition states of Bob's qubit depend on Alice's original qubit #2 through the initial entanglement in Step 1, as well as depending on the unknown qubit #1 to be teleported from the CNOT gate in Step 3. The reason we need to measure the state of Alice's qubit #2 and qubit #1 is to figure out the way Bob's qubit depends on these two. The current status is shown in Fig. 8.4. Note that Bob has not done anything with his qubit at this stage.

4. Alice now sends the two classical bits of information from the measurements to Bob by email or phone. According to Eq. (8.6), her measurements can be 00, 10, 01 or 11, each with 25% probability.

Depending on the measurement obtained by Alice, Bob can recover the original state of the teleported qubit (i.e., $a|0\rangle + b|1\rangle$) by using a combination of X or Z gates. The specific combination of X/Z gates to use will be explored as a question in Sect. 8.4. This situation is illustrated in Fig. 8.5. At this stage, the qubit has been successfully teleported from Alice to Bob, and thus ends the teleportation protocol.

Throughout the teleportation process, the original qubit #1 that has to be teleported does not remain in its original quantum state: $a|0\rangle + b|1\rangle$. This is because Alice performs a measurement on it during the teleportation protocol. As a result, there is no copy of qubit #1 existing at any time, and so teleportation does not

Fig. 8.4 Four possible superposition states of Bob's qubit

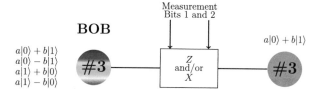

Fig. 8.5 The final result of teleportation between Bob and Alice

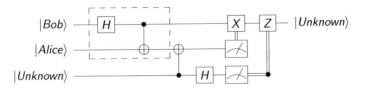

Fig. 8.6 The full quantum circuit for quantum teleportation. The dashed box entangles Alice's and Bob's qubits to make the Bell state. Afterwards, the quantum teleportation protocol described in the text is performed.

contradict the no-cloning theorem. It is important to understand that neither Alice nor Bob know what qubit #1's coefficients a or b are at any point in the process. All they know is that qubit #1 has been teleported from Alice to Bob. The full quantum teleportation circuit is illustrated in Fig. 8.6.

Why is this protocol interesting? To answer this, imagine Alice and Bob met a long time ago and each took one qubit of the entangled pair. Bob is now traveling around the world and can only communicate with Alice by phone or email. If Alice wanted to transfer quantum data to Bob without quantum teleportation, she would have to meet Bob and physically give Bob her qubit. Quantum teleportation allows Alice to send *quantum* information using a *classical* communications channel. All she has to do is make some measurements and email Bob the values. Bob can then apply the correct recipe to his qubit to bring it to the state of the original qubit #1. As well as sending information between two people, quantum teleportation is a useful way of causing interaction between different parts of a quantum computer (by teleporting a qubit to a different part of the quantum computer you want to interact with).[3]

8.3   Big Ideas

1. If Alice and Bob have a single qubit each, which they entangle, it is possible for Alice to teleport the information encoded in a third unknown qubit into Bob's qubit.
2. Quantum teleportation sends quantum information by using the entanglement and measurement properties of quantum mechanics.
3. Quantum teleportation does not destroy the qubit to be teleported (like in science fiction). It only transfers the information contained within the qubit without ever needing to know that information.

---

[3] Fermi National Accelerator Laboratory is building a quantum teleportation experiment which will extend over large distances, helping to develop a future quantum internet, e.g., https://qis.fnal.gov/quantum-teleportation-experiment/.

8.4   Check Your Understanding

1. ● Could quantum teleportation be used to teleport a physical object from one place to another? Why or why not?
2. ■ What would lead someone to think quantum teleportation can transmit information faster than the speed of light? Explain why this is not possible.
3. ■ By the no-cloning theorem, it is not possible to make a copy of an unknown qubit. At what point in the teleportation protocol does the unknown qubit collapse into a definite state?
4. ■ In the original protocol, Alice applies the CNOT and then measures Bit 2 (see Fig. 8.3). After this, Alice then applies the Hadamard to qubit #1 and then measures Bit 1 (see Fig. 8.3). What happens if she decides to reverse the procedure by measuring Bit 1 first, before applying the two-qubit CNOT gate?
5. ■ If Bob knows that his qubit is in the $b|0\rangle + a|1\rangle$ state, which gate(s) would he need to use to change it back into the original needed $a|0\rangle + b|1\rangle$ state?
    (A) X
    (B) Z
    (C) X then Z
6. ■ If Bob knows that his qubit is in the $a|0\rangle - b|1\rangle$ state, which gate(s) would he need to use to change it back into the original needed $a|0\rangle + b|1\rangle$ state?
    (A) X
    (B) Z
    (C) X then Z
7. ■ If Bob knows that his qubit is in the $a|1\rangle - b|0\rangle$ state, which gate(s) would he need to use to change it back into the original needed $a|0\rangle + b|1\rangle$ state?
    (A) X
    (B) Z
    (C) X then Z

Quantum Algorithms

9

We have come a long way from Chap. 1. To recap on what we have learnt, we have understood important quantum mechanical phenomena such as superposition and measurement (through the Stern-Gerlach and Mach-Zehnder experiments). We have also learnt that while quantum computers can in principle break classical encryption protocols, they can also be used to make new secure channels of communication. Furthermore, we have applied quantum logic gates to qubits to perform quantum computations. With entanglement, we teleported the information in an unknown qubit to another qubit. This is quite a substantial achievement.

However, we have not yet learned about a fundamental aspect of quantum computing: *quantum algorithms*. Simply put, given a task that we want the quantum computer to perform, a quantum algorithm is how the quantum computer performs this task on some input qubits. One typical example of an algorithm on a classical computer is the search algorithm, e.g., searching a database to find a friend in your friends list. In fact, quantum computers can also implement search algorithms. Grover's algorithm, one of the two most famous quantum computing algorithms (the other being Shor's algorithm which we learnt about in Chap. 5), uses entanglement to search a database faster than any classical computer can. While studying Grover's algorithm is outside the scope of this course, we will study the Deutsch-Jozsa Algorithm that shows how quantum computers can perform calculations faster than classical computers. After studying this algorithm, you will have a basis to learn more complicated algorithms.

9.1 ● The Power of Quantum Computing

The main advantage that quantum computers have over classical computers is **parallelism**. Because qubits can be in a superposition of states, a quantum computer can perform an operation on all of the states simultaneously. Let's say we want to know the result of applying some function $f(x)$ to some number x. Two classical

© The Author(s) 2021
C. Hughes et al., *Quantum Computing for the Quantum Curious*,
https://doi.org/10.1007/978-3-030-61601-4_9

computations are needed to find the result for $x = 0$ and for $x = 1$, whereas a quantum computer can evaluate both answers in parallel as displayed in Fig. 9.1.

If we wanted to compute $f(x)$ for $x = 2$ (represented as 10 in binary) and $x = 3$ (represented as 11), we would need to add a second qubit. The two-qubit quantum computer can then evaluate all four possibilities at once as shown in Fig. 9.2.

Question 1 How many pieces of information can a three-qubit quantum computer process in parallel? Write down all of the states.
The possible states are

$$|000\rangle, |001\rangle, |010\rangle, |100\rangle, |011\rangle, |110\rangle, |101\rangle, |111\rangle \rightarrow 8 \text{ pieces of information.}$$
(9.1)

Adding a qubit to a quantum computer doubles its processing power! For a classical computer, you need to double the number of wires in the processor to get double

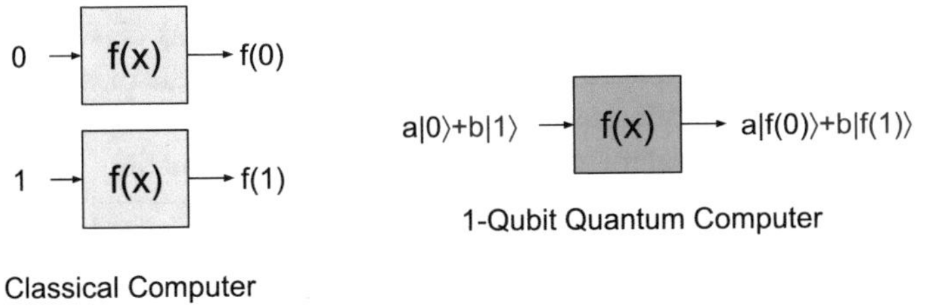

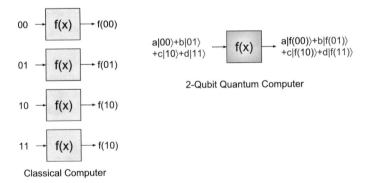

Fig. 9.1 It takes a classical computer two operations to operate on two pieces of information. A quantum computer with one qubit can operate on two classical pieces of information at once.

Fig. 9.2 It takes a classical computer four operations to operate on four pieces of information. A quantum computer with two-qubits can operate on four classical pieces of information at once.

the processing power.[1] However, with a quantum computer, you only need to add a single qubit to double the processing power! Further, an n-qubit system can perform certain 2^n operations at once!

Separate from the issue of processing power is a concept known as **memory**. In a classical computer, on a standard 64-bit laptop, each number can be represented in the 64-bit binary representation (a simple extension of the 8-bit binary representation you already learned about). If you wanted four numbers on a 64-bit machine at the same time, then you need to have $4 \times 64 = 256$-bits of memory on your hard drive to store them. On a 64 bit classical computer, for M different numbers, you need $M \times 64$-bits of memory; i.e., the bits needed for the memory is linear as a function of the number of numbers required. However, on an n-qubit quantum computer, there can be 2^n different coefficients of the quantum state that could in principle hold the numbers and therefore can be used as memory; i.e., the qubits needed for memory is logarithmic as a function of the number of numbers you want.

Because classical computers are very advanced and have large processing power and terabytes of memory, classical computers can simulate small quantum computers. As the addition of a single qubit would double the memory required, the largest supercomputer in the U.S.[2] would only be able to simulate a 46-qubit quantum computer. As of 2018, Google has a quantum computer with a quantum chip (called the Bristlecone) which has 72-qubits.

9.2 ● Limitations

While parallelism sounds amazing in theory, it is not immediately useful on its own. A quantum computation can calculate a superposition of the 2^n numbers, however a measurement still needs to be performed in order to extract information from the quantum computer. One measurement will only show one of those answers and afterwards collapse the superposition into a basis state. Think about it as if the 2^n numbers are all on a secret scratchpad that we cannot see, and nature shows you one random page at a time, then burns the scratchpad. You would need to run the quantum computer at least 2^n times to get all the numbers, therefore negating any advantage over classical computers. As an example of this, the two-qubit quantum computer can calculate the superposition $a|f(00)\rangle + b|f(01)\rangle + c|f(10)\rangle + d|f(11)\rangle$, but measuring this state will result in either $f(00)$, $f(01)$, $f(10)$, OR $f(11)$. If you are unlucky, due to the randomness of quantum physics, you could repeat the computation four times and still not see all of the possibilities.

Quantum computers are therefore only practical for certain types of problems. Since quantum computers are built on quantum physics principles, we intuitively

---

[1] It is an observation that classical computers double their processing power roughly every 18 months. This is known as Moore's law.

[2] The Titan at Oak Ridge Laboratory as of 2018.

expect that they would be best suited for simulating quantum phenomena directly. Generally, these types of problems look for correlations between different outputs. Due to this, it is generally accepted that quantum computers will not replace classical computers but will be able to perform different calculations that classical computers simply cannot. We will study an example problem which the quantum computer can solve more efficiently than a classical computer.

9.3   ◆ Deutsch-Jozsa Algorithm

Here we provide a proof that quantum computers can be faster than classical computers by explicit construction of a problem.

9.3.1   The Problem Statement

Let $f(x)$ be an unknown function that operates on a single qubit. There can only be four different functions that satisfy this requirement, and the four different functions are shown in Table 9.1.

A function is called **constant** if it always outputs the same result for all values of x. A function is called **balanced** if it outputs 1 for half of all the possible values of x and 0 for the other half. The question posed to the computer is this:

"Is the function $f(x)$ a constant function or a balanced function?"

For this single qubit case, the question is answered by checking if $f(0) = f(1)$. It also turns out in this single qubit case that there are only constant and balanced functions. However, in multiple qubit systems, there exist functions that are neither constant nor balanced. In the multiple qubit scenario, it is important that in the problem statement the function given to the quantum computer is either constant *or* balanced, and not something else.

Question 2 Which of the functions in Table 9.1 are constant and which are balanced?

The functions f_1 and f_4 are constant, while f_2 and f_3 are balanced.

Table 9.1 There are only four possible single qubit functions

f_1	f_2	f_3	f_4
$f_1(0) = 0$	$f_2(0) = 0$	$f_3(0) = 1$	$f_4(0) = 1$
$f_1(1) = 0$	$f_2(1) = 1$	$f_3(1) = 0$	$f_4(1) = 1$

Question 3 If you run the classical algorithm and see that $f(0) = 1$, could you tell whether the function is constant or balanced?

No, it could either be the balanced function f_3 or the constant function f_4. A classical computer would have to evaluate both $f(0)$ and $f(1)$ to determine the answer. How can a quantum computer determine the answer with only one measurement instead of two?

9.3.2 Conceptual Understanding

Before we go through the Deutsch-Jozsa Algorithm in detail, it will be useful to understand a cartoon solution of the problem, which we will demonstrate using the Mach-Zehnder interferometer from Chap. 3. Once again, superposition and interference will be the key properties to utilize. The cartoon experimental setup is shown in Fig. 9.3. In the QuVis simulation, we will model the functions by placing pieces of glass in the blue boxes. The goal is to illustrate how it may be possible to classify $f(x)$ as either constant or balanced by making a single measurement. Here is how the algorithm can be implemented:

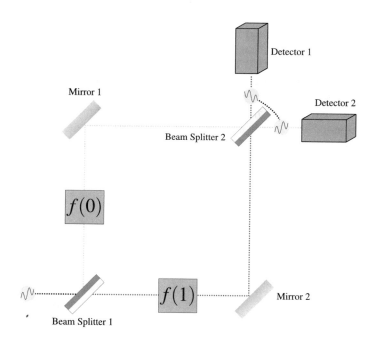

Fig. 9.3 The Mach-Zehnder interferometer altered to implement the cartoon version of the Deutsch-Jozsa algorithm. The function implementations are shown in Fig. 9.5.

Fig. 9.4 Inputs to the
function are photons along
two different paths. A photon
taking the yellow path is
$x = 0$, while a photon taking
the red path is $x = 1$.

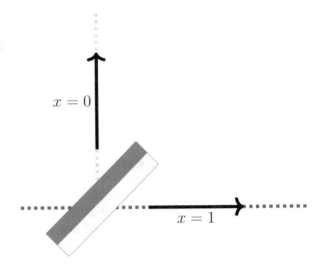

1. The two inputs $x = 0$ and $x = 1$ are represented by the two possible photon paths
   as shown in Fig. 9.4. A photon taking the yellow path is $x = 0$, while a photon
   taking the red path is $x = 1$. Beam splitter 1 therefore creates a superposition
   of 0 and 1 since the photon takes both paths. Due to the orientation of the beam
   splitter, the red transmitted path will have no phase shift whereas the yellow
   reflected path will have a phase shift of π.
2. Each of the four functions in Table 9.1 can be modelled by a different experimental setup as shown in Fig. 9.5. For example, if we wanted to test f_1, we would
   place a piece of glass along the red path but nothing along the yellow path. A
   photon passing through the glass will experience an additional phase shift of π.
   The reason that this is only a cartoon demonstration is that the phase shifters do
   not actually implement the function, as we will see in the next section.

   Question 4 If f_1 is being tested, what is the phase of the yellow path upon
   reaching the second beamsplitter? The red path photon?

   The yellow path was phase-shifted by Beam Splitter 1 and unaffected by the
   blue function box $f(0)$. The red path was unaffected by Beam Splitter 1 and
   phase-shifted by the blue function box $f(1)$. Therefore, they both have a phase
   shift of π.
3. The second beam splitter creates the interference necessary to ensure that
   measurement happens only in one detector. Depending on which detector is
   measured, this is interpreted as the function being constant or balanced.

   Question 5 For the experimental configuration f_1, what is the phase of the
   yellow path photon at Detector 1? The red path photon at Detector 1?

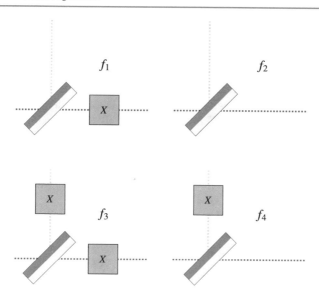

Fig. 9.5 The four different functions from Table 9.1 experimentally implemented by four different configurations. In this cartoon, we have denoted the function changing the bit by an X-gate, however in reality as described in Eq. (9.2) two qubits are needed to implement these functions.

At Detector 1, the yellow path and red path photons both have a phase shift of π.

Question 6 For the experimental configuration f_1, what is the phase of the yellow path photon at Detector 2? The red path photon at Detector 2?

At Detector 2, the yellow path photon has a phase shift of π while the red path photon has a phase shift of 2π.

4. Measure which detector is activated.

Question 7 For the experimental configuration f_1, which detector(s) go off and with what probability?

Detector 1 experiences constructive interference, while Detector 2 experiences destructive interference. Therefore, only Detector 1 activates for f_1, which is a constant function. Which detector(s) go off for f_2, f_3, and f_4?

After working through the exercises, you should see that thanks to superposition and interference, only one quantum measurement is needed in this cartoon picture of the Deutsch-Jozsa problem. The general algorithm is presented in the next section.

9.3.3   Quantum Algorithm

Before we describe the full quantum solution, we need to setup some useful tools. For example, in the quantum computing literature, it is common to use the mathematical tool called modular arithmetic. For this algorithm, we will not need to understand modular arithmetic more than basic notation. In quantum computing, modular arithmetic with "mod 2" is defined to be $f(0) \oplus f(1) = 0$ if $f(0) + f(1) = 0, 2, 4, 6, \ldots$. However, $f(0) \oplus f(1) = 1$ if $f(0) + f(1) = 1, 3, 5, \ldots$. Note the circle with a plus inside $\oplus$ denotes this modular arithmetic "mod 2" operation. The $\oplus$ operation outputs the remainder of dividing a number x by the number 2. As an example, if $f(0) = 0$ and $f(1) = 1$ then $f(0) \oplus f(1) = 1$, whereas if $f(0) = 1$ and $f(1) = 1$ then $f(0) \oplus f(1) = 0$.

Also, we will need a second qubit for this algorithm, and will shortly see why. In the quantum computing world, the function $f(x)$ is implemented by

$$|x\rangle|y\rangle \xrightarrow{f} |x\rangle|y \oplus f(x)\rangle. \qquad (9.2)$$

As an example, assume $f(0) = 1$, then $|0\rangle|1\rangle \xrightarrow{f} |0\rangle|1 \oplus f(0)\rangle = |0\rangle|0\rangle$. Although the implementation of functions as in Eq. (9.2) looks strange, this is needed to ensure that the function operation is unitary.[3] The circuit that implements the Deutsch-Jozsa algorithm is shown in Fig. 9.6. We will now give a walk-through of the algorithm and the circuit.

Deutsch-Jozsa Procedure:

1. As the first step of the algorithm shown in Fig. 9.6, get two qubits, and put them into a $|0\rangle|1\rangle$ product state. In the modified Mach-Zehnder experiment above, only the first qubit from Fig. 9.6 was shown. The second qubit was hidden in the blue function boxes.
2. Operate on each qubit with the Hadamard gate. Following the rules of the Hadamard gate, the two qubit state is now

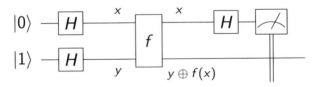

Fig. 9.6 The quantum circuit for the one qubit Deutsch-Jozsa algorithm. The generic function $f(x)$ is represented by the box with f inside, and the labels below/above the lines indicate how the function is implemented.

---

[3]Taking the function f_1 and representing its action on a single qubit as a matrix, this matrix is not unitary. Non-unitarity violates the laws of quantum mechanics.

$$\frac{1}{2}(|0\rangle + |1\rangle)(|0\rangle - |1\rangle). \tag{9.3}$$

In the Mach-Zehnder cartoon in Fig. 9.3, Beam splitter 1 performs the first Hadamard gate on the first qubit in Fig. 9.6.

3. Apply the function $f(x)$ using the rule in Eq. (9.2) to the state in Eq. (9.3). After performing the arithmetic, the two qubit state can be organised as

$$\frac{1}{2}\left(|0\rangle\left(|0 \oplus f(0)\rangle - |1 \oplus f(0)\rangle\right) + |1\rangle\left(|0 \oplus f(1)\rangle - |1 \oplus f(1)\rangle\right)\right). \tag{9.4}$$

In order to get a clearer picture of the effect of $f(x)$ on the state, it is useful to notice that if $f(0) = 0$ then $|0 \oplus f(0)\rangle - |1 \oplus f(0)\rangle = |0\rangle - |1\rangle$. Additionally, if $f(0) = 1$ then $|0 \oplus f(0)\rangle - |1 \oplus f(0)\rangle = -|0\rangle + |1\rangle$. We can combine these two by writing $|0 \oplus f(0)\rangle - |1 \oplus f(0)\rangle = (-1)^{f(0)}(|0\rangle - |1\rangle)$. A similar formula is needed for the $f(1)$ case also, and we leave this as an exercise for the reader. Applying this formula to Eq. (9.4) gives

$$\frac{1}{2}\left((-1)^{f(0)}|0\rangle\left(|0\rangle - |1\rangle\right) + (-1)^{f(1)}|1\rangle\left(|0\rangle - |1\rangle\right)\right) \tag{9.5}$$

$$= (-1)^{f(0)}\frac{1}{2}\left(|0\rangle + (-1)^{(f(0)+f(1))}|1\rangle\right)(|0\rangle - |1\rangle). \tag{9.6}$$

In the Mach-Zehnder cartoon in Fig. 9.3, the interaction between the two qubits was modeled by the photon passing through the blue function boxes.

4. We now throw away the second qubit. We only keep the first qubit and make sure it is normalized correctly. The first qubit is

$$\frac{1}{\sqrt{2}}(|0\rangle + (-1)^{(f(0)+f(1))}|1\rangle). \tag{9.7}$$

The reason we need the second qubit is to perform the gate operations and collect the like-terms which ensures the algorithm works. This second qubit is called an ancilla qubit because it is not measured. This is shown in the circuit in Fig. 9.6 as the lack of the measurement operator in the second qubit line.

5. Apply a Hadamard gate to the qubit state in Eq. (9.7) to produce

$$\frac{1}{2}\left(\left(1 + (-1)^{(f(0)+f(1))}\right)|0\rangle + \left(1 - (-1)^{(f(0)+f(1))}\right)|1\rangle\right). \tag{9.8}$$

In the Mach-Zehnder cartoon in Fig. 9.3, the second Hadamard gate operation was implemented by Beam Splitter 2.

6. Measure the qubit. If $f(x)$ is constant, then the state in Eq. (9.8) reduces to $|0\rangle$, while if $f(x)$ is balanced then the state reduces to $|1\rangle$. In the Mach-Zehnder cartoon in Fig. 9.3, the detector measured the final state of the photon.

As this algorithm shows, a single measurement of $|0\rangle$ or $|1\rangle$ shows whether the function is constant or balanced. Impressively, this algorithm straightforwardly extends to functions that take in any number of inputs. This is impressive because only *one single* measurement can tell you whether a function of any size is constant or balanced. For a classical computer to do the same task, it would need to measure each of the inputs, which is exponentially slower.

9.4   ● Quantum Computers Today

While the Deutsch-Jozua problem has no known commercial applications, useful quantum algorithms such as Shor's factoring algorithm rely upon similar concepts. Quantum algorithms are believed to exist that can speed up machine learning algorithms and efficiently simulate the quantum behavior of molecules. As of 2018, companies such as IBM and Google have built different types of quantum computers that contain up to 72 qubits. To give you an idea of where we need quantum computers to be, factoring a 1024-bit modern encryption key using Shor's algorithm would require more than 5,000 qubits. In 2019, Google claimed to have performed[4,5] the first quantum computation that a classical computer could not do—a milestone known as "quantum supremacy". Quantum supremacy means that a quantum computer can solve a problem that a classical computer cannot. However, the solution of the problem may not be of practical use. As such, it is important to note that Google has demonstrated quantum supremacy, not the "quantum usefulness" milestone. Google performed their task on a 53-qubit quantum computer, which took 200 s. They claimed it would take a classical computer 10,000 years to do the same task. However, shortly after, IBM suggested[6] that an improved classical supercomputing technique could theoretically perform the task in just 2.5 days.

Different technological difficulties may be encountered when improving a quantum computer. As we have mentioned, a quantum computer can be built using lasers.[7] However, there are also random photons outside of the quantum computer in the environment that may accidentally leak into the quantum computer, and these environmental photons can then cause accidental changes to the quantum state. Such accidental changes are called "noise". To reduce the number of these environmental photons, the quantum computer needs to be be cooled down to near absolute zero (around $-450°$ Fahrenheit). However, this is difficult. The more qubits you add, the more you need to keep at this low temperature (a technological challenge). Also, the more qubits you add, the more lasers you need

---

[4]https://www.nature.com/articles/s41586-019-1666-5.

[5]https://www.sciencenews.org/article/google-quantum-supremacy-claim-controversy-top-science-stories-2019-yir.

[6]https://arxiv.org/abs/1910.09534.

[7]A laser is a source of photons which have the same wavelength and are in phase.

to interact the qubits. It is technologically difficult to keep lots of qubits in one small space, but also cause isolated interactions between them using different lasers. Further, the more qubits you add, the more likely it is that the qubits will interact accidentally with the environment, which will then destroy the system's quantum properties through a process known as decoherence. However, given how classical computers went from being the size of a room in the 1960s to an iPhone within a few decades, governments and industries are investing billions of dollars towards making quantum computers realistic. Ultimately, quantum computers are destined to complement classical computers, not replace them, so don't expect to have a quantum phone in your pocket anytime soon![8]

9.5   Big Ideas

1. Quantum computers can perform a function operation on all (qu)bits simultaneously - which is called parallelism. This is an advantage over classical computers.
2. Getting the results from the quantum computation requires measuring the qubits. Too many measurements could ruin the quantum advantage.
3. The Deutsch-Jozsa Algorithm solves a toy problem on a quantum computer faster than a classical computer can.

9.6   Activities

◆ Explore more quantum algorithms from the IBM quantum textbook.[9]

9.7   Check Your Understanding

1. ●
   (a) How many different classical pieces of information can be represented by eight classical bits (1 byte)?
   (b) What about a quantum computer with eight qubits?
   (c) What advantage does the quantum computer have over the classical computer?
2. ◆ This problem refers to the experimental setup in Fig. 9.3. Which detector(s) go off for the function
   (a) f_1?

---

[8]Theoretical physicists and computational scientists at Fermi National Accelerator Laboratory are working on improving algorithms and the foundations of quantum science in order to expand the problems that (near term) quantum devices can solve, e.g., https://qis.fnal.gov/quantum-computing-for-hep/.

[9]https://qiskit.org/textbook/ch-algorithms/index.html.

$$f(x)$$

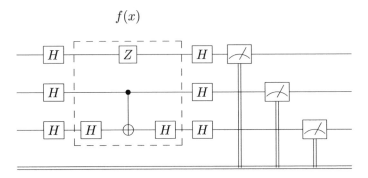

Fig. 9.7 The gate implementation for testing the different possible three-qubit functions.

   (b) f_2?
   (c) f_3?
   (d) f_4?

3. ◆
   (a) Which detector(s) go off if the function is constant?
   (b) Which detector(s) go off if the function is balanced?
   (c) How many photons would you need to send to determine whether the function was constant or balanced?

4. ◆ Explain how superposition and interference allows the Deutsch-Jozsa algorithm to beat the classical algorithm.

5. ◆ Figure 9.7 shows the gate implementation for testing a three-qubit function $f(x)$. A constant function will always result in $|000\rangle$ or $|111\rangle$.
   (a) How many evaluations would be needed on a classical computer to tell whether this function is constant or balanced?
   (b) By running this algorithm on IBM Q, can you determine whether this function is constant or balanced?

Worksheets

10

10.1 ◆ Correlation in Entangled States Lab

Objectives:

- Experimentally determine the difference between two particles in a product state vs. an entangled state using the entanglement simulator.[1]
- Apply the idea of basis changing to explain the correlation that is observed.

Questions

Alice and Bob each measure one of two qubits with a Stern-Gerlach apparatus. Start with both SGAs along the z-axis (Fig. 10.1)

1. Try sending pairs of particles in a product state $|\uparrow_A\rangle|\downarrow_B\rangle$. What do Alice and Bob measure individually?
2. Try sending pairs of particles in an entangled state: $\frac{1}{\sqrt{2}}(|\uparrow_A\rangle|\downarrow_B\rangle - |\downarrow_A\rangle|\uparrow_B\rangle)$. What do Alice and Bob measure individually?
3. If Alice measures her spin, would you be able to predict Bob's result:
   (a) In the product state?
   (b) In the entangled state?

   Now rotate both SGAs along the x-axis (Fig. 10.2).

4. Try sending pairs of particles in a product state $|\uparrow_A\rangle|\downarrow_B\rangle$. What do Alice and Bob measure individually?

---

[1]https://www.st-andrews.ac.uk/physics/quvis/simulations_html5/sims/entanglement/entanglement.html.

© The Author(s) 2021
C. Hughes et al., *Quantum Computing for the Quantum Curious*,
https://doi.org/10.1007/978-3-030-61601-4_10

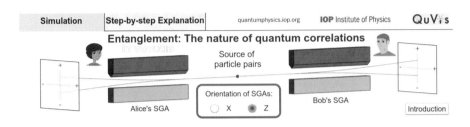

Fig. 10.1 Figure reproduced from the QuVis website, licensed under creative commons CC-BY-NC-SA.

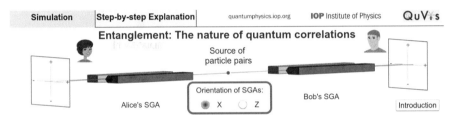

Fig. 10.2 Figure reproduced from the QuVis website, licensed under creative commons CC-BY-NC-SA.

5. Try sending pairs of particles in an entangled state $\frac{1}{\sqrt{2}}(|\uparrow_A\rangle|\downarrow_B\rangle - |\downarrow_A\rangle|\uparrow_B\rangle)$. What do Alice and Bob measure individually?

6. If Alice measures her spin, would you be able to predict Bob's result:
   (a) In the product state?
   (b) In the entangled state?

7. Convert the product state $|\uparrow_A\rangle|\downarrow_B\rangle$ into the x-basis and use it to explain the observations in the x-basis. Recall that $|\uparrow\rangle = \frac{1}{\sqrt{2}}(|+\rangle + |-\rangle)$ and $|\downarrow\rangle = \frac{1}{\sqrt{2}}(|+\rangle - |-\rangle)$.

8. Convert the entangled state $\frac{1}{\sqrt{2}}(|\uparrow_A\rangle|\downarrow_B\rangle - |\downarrow_A\rangle|\uparrow_B\rangle)$ into the x-basis and use it to explain the measurements in the x-basis.

9. Suppose that there are two possible sources of particles. Source #1 randomly emits two particles in either the state $|\uparrow_A\rangle|\downarrow_B\rangle$ or $|\downarrow_A\rangle|\uparrow_B\rangle$ with equal probability. Source #2 emits two particles in the entangled state $\frac{1}{\sqrt{2}}(|\uparrow_A\rangle|\downarrow_B\rangle - |\downarrow_A\rangle|\uparrow_B\rangle)$. How can Alice and Bob tell whether the source is #1 or #2?

10.2 ■ Polarizer Demo

For students who have learned about polarization, the creation of superposition states can be demonstrated using three polarizing filters. When unpolarized light is sent through a vertical filter, only vertically polarized light is able to pass through. Sending vertically polarized light through a horizontal filter results in no light passing through, since the vertical and horizontal polarizations are mutually exclusive. Surprisingly, adding a diagonal filter in between recovers the light! The diagonal polarizer introduced a horizontally polarized component, similar to how passing a spin-up electron through a horizontal SGA created a horizontal superposition.

Question Relate the behavior of the polarizers to what you saw in the SGAs. Hint: think of the top two polarizers in Fig. 10.3 as the z-basis, and diagonal polarizers as the x-basis.

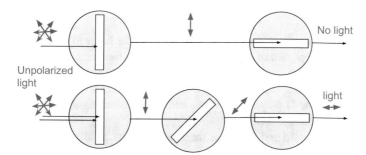

Fig. 10.3 Unpolarized light is sent through a series of polarizing filters.

10.3   ● Quantum Tic-Tac-Toe

Quantum Tic-Tac-Toe was developed by Alan Goff in 2004 as a metaphor to teach quantum concepts such as superposition, entanglement, and measurement collapse. It has been found to be a helpful strategy in teaching quantum mechanics to undergraduate students at Purdue, especially for students who struggle with grasping the concepts.[2]

Quantum Tic-Tac-Toe resembles the classical Tic-Tac-Toe game in its setup and objective of completing three in a row. However, the game uses characteristics of quantum systems, so instead of using one marker X or O, the players use pairs of Xs and Os, which are traditionally called "spooky," after Einstein's reference to entanglement as "spooky action at a distance".[3] Using indices for each marker's move is important when determining the winner of the game. Additionally, we use a color code for each player and connect the spooky markers to help students better visualize the game process. We also number the squares for future reference.

10.3.1 The Rules

1. The X player goes first. We note that keeping indices helps to track the game. The markers can be placed in any of the two spaces on the game board (Fig. 10.4).
2. The O player goes next. The markers can be placed in any two squares, even ones that are already occupied by other X or O markers. Notice in Fig. 10.5 that the index for the O player also starts with 1, representing its first move placing markers in squares 1 and 6.
3. Player X goes again and can place their spooky markers at any two squares, even ones occupied by other Xs or Os. The game goes on until the players create a "cyclic loop" as seen in Fig. 10.6.
4. **Collapsing the quantum state**. When a loop is created, the players have to collapse their state. There are three options for who makes the decision on how

Fig. 10.4 The Quantum Tic-Tac-Toe layout with numbered squares (left): one player's move with spooky markers x_1 (right).

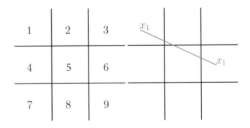

[2]Hoehn R, et al. (2014). "Using Quantum Games to teacher quantum mechanics, Part 1." *Journal of Chemical Education 91* (3), 417–422. Retrieved from https://pubs.acs.org/doi/ipdf/10.1021/ed400385k.

[3]Einstein, Podolsky, and Rosen (1935) "Can quantum-mechanical description of physical reality be considered complete?" *Physical Review, 47*: 777–780. Retrieved from https://journals.aps.org/pr/pdf/10.1103/PhysRev.47.777.

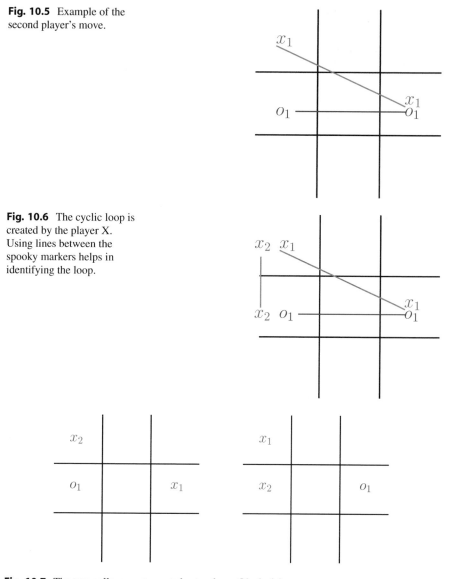

Fig. 10.5 Example of the second player's move.

Fig. 10.6 The cyclic loop is created by the player X. Using lines between the spooky markers helps in identifying the loop.

Fig. 10.7 The two collapse outcomes due to player O's decision.

the markers will be collapsed. The fair choice would be by the player who did not create the cycle (in this case, player O). When the markers are forced to collapse, only one of the two squares for each move can be chosen, so player O can choose either square 4 or 6. Depending on their choice, the outcome would be different (Fig. 10.7). Once the states are collapsed, the "spooky markers" change into classical markers and they fully occupy the state of one particular square.

Fig. 10.8 Player X wins, because the sum of their indexes is $1 + 2 + 3 = 6$. Player O got three in a row, but the sum of their indexes is $2 + 1 + 4 = 7$.

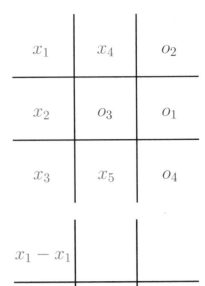

Fig. 10.9 A player cannot put both markers in the same square.

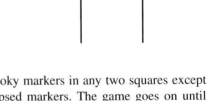

5. The next player can place his or her spooky markers in any two squares except the ones that are occupied by the collapsed markers. The game goes on until another cycle is created and the players are forced to collapse the state.

6. **Winning the game.** In some cases both players will create three in a row after collapsing their spooky markers. In this case, the player with the smallest sum of indexes wins. For example, in Fig. 10.8 player X wins because they have the smaller sum.

Some Other Rules Can Be Added or Modified One of the requirements could be that players cannot place both markers in the same square like the one shown in Fig. 10.9. Another way to make the collapse more quantum (or more random) is using a coin flip to decide which player chooses the collapse.

Other modifications may include assigning different point values for three in a row, such as the winner with lowest sum of the indexes gets 1 point, while the other player gets 1/2 point.

One of the main challenges of playing the game is to observe when a cycle has been created so the state of the spooky markers can be collapsed at the right time.

A computer-simulated game will automatically keep track of this and will force students to collapse their markers, such as this game simulator.[4]

We found that using color codes and connecting lines helps visually track loops. Another way is to create a model of the game where students can see the connections and collapse the states using physical pieces. It would be interesting to see students' responses as to which medium helps them understand the game principle better.

10.3.2 Connection to Quantum Physics

How are the game rules and principles connected to the real applications of quantum mechanics? There are three major themes that can be drawn from the game: superposition, the effect of measurement, and entanglement.

Superposition

In classical physics all objects have defined states. However, quantum systems can exist in a superposition of several classical states at the same time. The example could be an electron with a spin that is in superposition of up and down, or a photon in a superposition of vertical and horizontal polarization. QTTT spooky markers exist in two separate locations on the game board, representing their state as a superposition state of two classical TTT markers.

Measurement

When measuring the state of a quantum system, the quantum state of a system collapses and only one classical state is observed with some probability. In QTTT, the rule of creating the loop forces players to collapse their markers (measure their quantum state). In this case the player decides how to collapse the markers, which corresponds to the scientist choosing the way of measuring quantum system, such as axis orientation. The rule of forcing the measurement when the loop is created does not have an exact corresponding physical meaning. Quantum systems can exist in a superposition state for an extended time, and the measurement is not forced, but chosen by the observer.

Entanglement

Entanglement is the quantum phenomenon of creating two or more particles, whose states cannot be described separately, but have some correlation even when they are separated by a significant distance. When measuring the state of one of the entangled particles, the state of the other particle can be known even without measurement. Einstein called it "spooky action at a distance." When the players collapse their states after creating a loop in QTTT, they know for sure in which state each marker would collapse into.

---

[4]http://qttt.rohanp.xyz/.

10.4   ● Schrödinger's Worm Using Five Qubits

Objectives

Design, build, and test quantum circuits that model systems in superposition and entanglement.

Setup

Open the IBM Q simulator[5] and start a new circuit in the Circuit Composer (Fig. 10.10).

The default is 5-qubits initialized to the $|0\rangle$ state. Gates can be applied by dragging and dropping them onto the appropriate qubit(s). Don't forget to add the measurement gate at the end to see the results. When you are satisfied with your circuit, save the experiment and click Run (Fig. 10.11).

By default, the circuit will be evaluated 1024 times using the simulator backend. You may also run the circuit on a real quantum computer, subject to a waiting period. Increasing the number of shots will increase the statistical accuracy of the results at the expense of run-time. After you have run the circuit, the results will appear in a link at the bottom of the page.

Fig. 10.10 A new experiment on the IBM Q Circuit Composer. Reprint courtesy of International Business Machines Corporation, ©International Business Machines Corporation.

---

[5]https://quantum-computing.ibm.com.

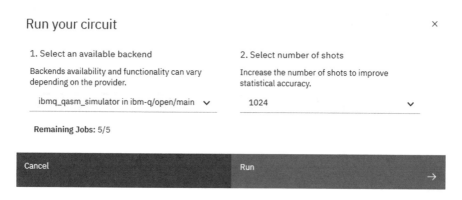

Fig. 10.11 Options for running the IBM Q experiment. Reprint courtesy of International Business Machines Corporation, ©International Business Machines Corporation.

Fig. 10.12 Dead or alive worms.

Part I: Superposition

The worm is alive when all five squares are black and dead when only four are black. Use a 0 to represent a white square and 1 to represent a black square (Fig. 10.12).

1. What is the classical state of the live 5-bit worm?
2. What is the classical state of the dead 5-bit worm?
3. Use IBM Q to create a worm in a superposition state of alive and dead. Let q[0] correspond to the bit on the far right.
4. Run the simulation and interpret the histogram.
5. How can you modify the circuit so that the worm is first put in a superposition state and then brought to life?
6. How can you modify the circuit so that the worm in a superposition state becomes definitely dead?

Fig. 10.13 Very dead or
alive worms.

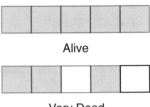

Alive

Very Dead

Part II: Entanglement

The worm is next to a hungry bird, such that the worm is either alive or chomped to
pieces (Fig. 10.13).

7. What is the classical state of the very dead worm?
8. Create a circuit that produces a worm in a superposition state of alive and very
   dead. (Hint: Two of the qubits are entangled.)
9. Run the simulation and interpret the histogram.
10. How can you modify the circuit so that the worm in a superposition state
    becomes either definitely dead or definitely alive?

Further Resources

- The Qiskit webpage[6] has resources for YouTube videos and other educational
  links.
- IBM quantum offers a (virtual) summer school[7] with minimal prerequisites
  required.

---

[6]https://qiskit.org/.
[7]https://qiskit.org/events/summer-school/.

10.5 ◆ Superposition vs. Mixed States Lab

Objectives
- Experimentally determine the difference between particles in a **superposition state** and a **mixed state** using the superposition states and mixed states simulator.[8]
- Apply the idea of basis changing to explain the experimental results.
- Compute the probability amplitudes given measurement results.

Questions
1. We send 100 electrons of unknown spin into a Stern-Gerlach apparatus. We measure that 50 are spin up and 50 are spin down. We can conclude that:
   (a) 100 electrons were in a 50/50 superposition state of up and down (superposition state).
   (b) The electrons were a classical mixture of 50 electrons spin up and 50 spin down (mixed state).
   (c) Not enough information
2. Use the simulator (Fig. 10.14) to compare the measurement outcomes of the mixed particles vs. the superposition particles. What are the similarities and differences?
3. By making a basis change with $|0\rangle = \frac{1}{\sqrt{2}}|+\rangle + \frac{1}{\sqrt{2}}|-\rangle$ and $|1\rangle = \frac{1}{\sqrt{2}}|+\rangle - \frac{1}{\sqrt{2}}|-\rangle$, can you explain the similarities and differences mathematically?
4. Which of the two inputs labelled "Superposition or mixture?" and "Superposition or mixture??" is a random mixture and which is a superposition?
5. The mixture consists of a fraction A of spin up particles and a fraction B of spin down particles. Find these fractions, A and B.
6. The superposition state can be written as $\alpha|0\rangle + \beta|1\rangle$. Find the amplitudes α and β assuming they are real and positive.
7. Use a basis change to show that the amplitudes α and β give the correct probabilities in both the x- and z- basis.

---

[8]https://www.st-andrews.ac.uk/physics/quvis/simulations_html5/sims/superposition/superposition-mixed-states.html.

Superposition states and mixed states

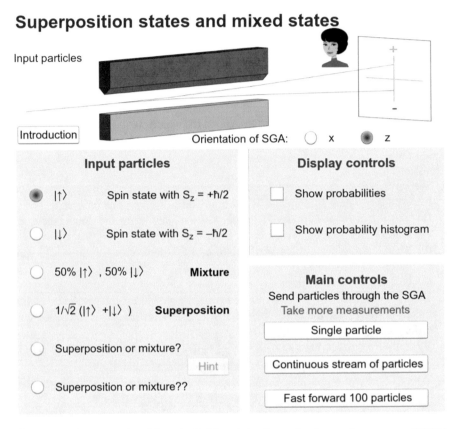

10.6 ◆ Measurement Basis Lab

Objectives

- Use the PHET Stern-Gerlach Simulator[9] to see how changing the orientation of the Stern-Gerlach Apparatus (SGA) affects the spin measurement.
- Perform calculations to write the spin in a different measurement basis (Fig. 10.15).

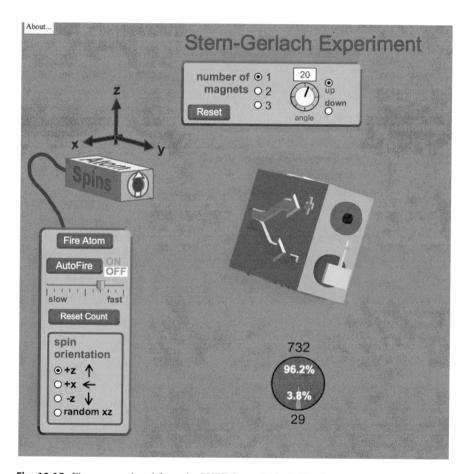

Fig. 10.15 Figure reproduced from the PHET Stern-Gerlach Simulator website, licensed under creative commons CC-BY.

---

[9]https://phet.colorado.edu/sims/stern-gerlach/stern-gerlach_en.html.

Angle of SGA (θ_{SGA})	Probability of going through	Probability of being blocked
0°		
15°		
30°		
45°		
60°		
75°		
90°		
105°		
120°		
135°		
150°		
165°		
180°		

Questions

1. Send spin up electrons through a single SGA and record the measurement probabilities for different SGA angles (see above table).
2. Generate a scatter plot of the data.
3. What function describes the shape of the graph?
4. Write the state of the spin up electron as a superposition for an arbitrary SGA angle (θ_{SGA}). In other words, find α and β in $|$electron$\rangle = \alpha|$goes through$\rangle + \beta|$blocked$\rangle$. The diagram below may help, but note that $\theta \neq \theta_{SGA}$.

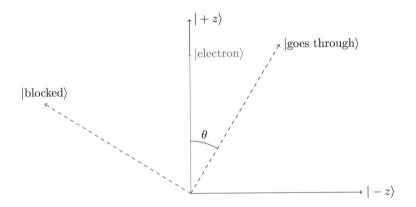

5. Do the theoretical probabilities match the simulated data?
6. What would your scatter plot look like if you sent electrons through with the random xz spin option?
7. What is the theoretical probability of spin down electrons passing through a SGA angled at 45°?
8. What is the theoretical probability of spin $+x$ electrons passing through a SGA angled at 45°?

10.7 ● One-Time Pad

10.7.1 One-Time Pad: Alice

Before parting ways, you and Bob agree on a key. Using a coin with heads = 0 and tails = 1, randomly generate a key of the same length as the message. Make sure that you and Bob have the same key.

- Shared Key:

Table 10.1 One-time pad (Alice)

Character	Binary code
A	01000001
B	01000010
C	01000011
D	01000100
E	01000101
F	01000110
G	01000111
H	01001000
I	01001001
J	01001010
K	01001011
L	01001100
M	01001101
N	01001110
O	01001111
P	01010000
Q	01010001
R	01010010
S	01010011
T	01010100
U	01010101
V	01010110
W	01010111
X	01011000
Y	01011001
Z	01011010

- Encoding:

    1. Choose a secret letter to send to Bob in binary using Table 10.1. Message:

    2. Add the key to your message, bit by bit, to encode the message. In binary,
       $0+0 = 0, 0+1 = 1+0 = 1$, and $1+1 = 0$. For example, if the key $= 0110$
       and the message $= 1101$, then the cipher text $= 1011$, as $0110+1101 = 1011$.
       – Cipher Text:

    3. Send the cipher text to Bob.

- Decoding

    1. Write down the cipher received from Bob.

Cipher from Bob							
Shared Key							

    2. Add the key to Bob's message, bit by bit, to decode the message.

Decoded message							

    3. What was the message?

- Eavesdropping

    1. Swap cipher texts with another group. How could you recover the original
       message?
    2. How many different keys would you need to try?
    3. If the original message had five letters instead of one letter, how many
       different keys would you need to try?
    4. You intercept a five letter message and, by chance, find a key that decrypts it
       to read HELLO. What other words could it possibly be?

• <u>Questions</u>

1. Why does adding the key to the cipher recover the original message?
2. Why is the one-time pad theoretically unbreakable?
3. What is the practical security flaw in the one-time pad?

10.7.2 One-Time Pad (Bob)

Before parting ways, you and Alice agree on a key. Using a coin with heads = 0 and tails = 1, randomly generate a key of the same length as the message. Make sure that you and Alice have the same key.

- Shared Key:

- Encoding:

    1. Choose a secret letter to send to Alice in binary. (Table 10.2) Message:

    2. Add the key to your message, bit by bit, to encode the message. In binary, $0 + 0 = 0, 0 + 1 = 1 + 0 = 1$, and $1 + 1 = 0$. For example, if the key = 0110 and the message = 1101, then the cipher text = 1011. $0110 + 1101 = 1011$.
        - Cipher Text:

Table 10.2 One-time pad (Bob)

Character	Binary code
A	01000001
B	01000010
C	01000011
D	01000100
E	01000101
F	01000110
G	01000111
H	01001000
I	01001001
J	01001010
K	01001011
L	01001100
M	01001101
N	01001110
O	01001111
P	01010000
Q	01010001
R	01010010
S	01010011
T	01010100
U	01010101
V	01010110
W	01010111
X	01011000
Y	01011001
Z	01011010

3. Send the cipher text to Alice.

- Decoding

    1. Write down the cipher received from Alice.

Cipher from Alice							
Shared Key							

2. Add the key to Alice's message, bit by bit, to decode the message.

Decoded message							

3. What was the message?

- Eavesdropping

    1. Swap cipher texts with another group. How could you recover the original message?
    2. How many different keys would you need to try?
    3. If the original message had five letters instead of one letter, how many different keys would you need to try?
    4. You intercept a five-letter message and, by chance, find a key that decrypts it to read HELLO. What other words could it possibly be?

- Questions

    1. Why does adding the key to the cipher recover the original message?
    2. Why is the one-time pad theoretically unbreakable?
    3. What is the practical security flaw in the one-time pad?

10.8 ⬤ BB84 Quantum Key Distribution

10.8.1 BB84 Quantum Key Distribution: Alice

- No Eavesdropper

    1. Randomly choose to prepare the electron in either the x- or z-basis.
    2. The electron that's sent through your Stern-Gerlach apparatus will either be in a 0 or 1 state. You can randomize this by flipping a coin.
    3. Pass the correct spin card to Bob face down.

$$\uparrow \; = 0 \qquad \downarrow \; = 1$$

$$\longleftarrow = 0 \qquad \longrightarrow = 1$$

    4. Once you have filled up the chart, tell Bob the basis used for each bit. If Bob tells you to "discard" the bit, cross it out on your chart.
    5. Check to see that you and Bob end up with the same sifted key.

Basis: x or z								
Bit value: 0 or 1								

- SIFTED KEY: _____
- With Eavesdropper

   1. Repeat the procedure, but instead of passing the spin card directly to Bob, it will first pass through Eve.
   2. Compare the sifted key bits one at a time. How can you tell if Eve intercepted the message?

- SIFTED KEY: _____

Basis: x or z							
Bit value: 0 or 1							

10.8.2 BB84 Quantum Key Distribution: Bob

- No Eavesdropper

1. Randomly choose between the x- or z-basis.
2. Receive the spin card from Alice and flip it over.
   - If your basis is the same as the card's, record the bit value.
   - If your basis is different, the output of your Stern-Gerlach apparatus will be random. Randomly pick 0 or 1.

$$\uparrow \ = 0 \qquad \downarrow \ = 1$$

$$\longleftarrow \ = 0 \qquad \longrightarrow \ = 1$$

3. Once you have filled up the chart, Alice will tell you the basis used for each bit. If you measured in a different basis, tell Alice to "discard" the bit and cross it out on your chart.
4. Check to see that you and Alice end up with the same sifted key.

Basis: x or z								
Bit value: 0 or 1								

- SIFTED KEY: _____
- <u>With Eavesdropper</u>

    1. Repeat the procedure, but instead of getting the spin card directly from Alice, it will first pass through Eve.
    2. Compare the sifted key bits one at a time. How can you tell if Eve intercepted the message?

- SIFTED KEY: _____

Basis: x or z							
Bit value: 0 or 1							

10.8.3 BB84 Quantum Key Distribution: Eve

- <u>With Eavesdropper (You!)</u>

  1. Randomly choose between the x- or z-basis.
  2. Receive the spin card from Alice and flip it over.
     - If your basis is the same as the card's, record the bit value and pass it along to Bob.
     - If your basis is different, the output of your Stern-Gerlach apparatus will be random. Randomly pick 0 or 1 for your bit value, erase Alice's value, write yours on the card, and pass the card along to Bob.
  3. Listen in as Alice and Bob compare their basis. If Bob says to "discard" the bit, cross it out on your chart.
  4. Compare your sifted key to Alice and Bob's key. Was your eavesdropping successful?

- SIFTED KEY: _____

Basis: x or z							
Bit value: 0 or 1							

Probability

A

A.1   Introduction

Probability is the branch of mathematics needed to understand how likely certain events are to occur. The study of probability appears all over in everyday life. Some examples of everyday probabilities include the chance you get heads when you flip a coin, or the likelihood it will rain tomorrow.

More specifically, an **event** is a group of one or more outcomes that form an observation, i.e., a coin flip. To find the probability associated with a specific outcome of an event, it is often necessary to perform many repeated experiments of the event. Each event in this set is referred to as a **trial**. The distribution of trials will follow the probability for the given event to occur. For example, Fig. A.1 shows the probability for a coin to land on heads a specific number of times after ten coin flips. Later in this appendix, we will work out the probabilities when there are five coin flips.

The set of all possible outcomes is often referred to as the **sample space**. Some examples of sample spaces are:

- Heads and tails when flipping a coin
- The different numbers on a die
- The 52 cards in a deck of cards

A.2   Combining Probabilities

With the above knowledge, we can now calculate the probabilities for more complicated scenarios. For example, by the end of this section, we will be able to find the probabilities for a specific number of heads to occur after we flip 5 coins.

When combining the probability for a specific event with that for another event, we can start by listing all the possible outcomes. For example, consider rolling a die

© The Author(s) 2021
C. Hughes et al., *Quantum Computing for the Quantum Curious*,
https://doi.org/10.1007/978-3-030-61601-4

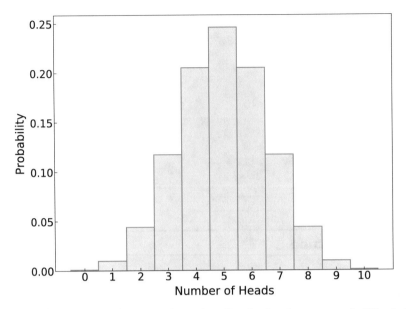

Fig. A.1 The chances of flipping a specific number of heads when there are a total of 10 coin flips

and flipping a coin; the possible outcomes are shown in Fig. A.2. The total number of unique outcomes is 12, and all of the different outcomes occur exactly once. From this, we can concluded that the probability of any one of the events listed above is $\frac{1}{12}$. We can also find this result mathematically from the probability of rolling a given number on a die, and separately the probability of getting heads or tails when flipping a coin. To find the combined probability for a given number on the dice (lets say a 4) and flipping a heads on a coin, we can multiply the two separate probabilities together to obtain the probability for both to occur (i.e., $\frac{1}{6} \times \frac{1}{2} = \frac{1}{12}$), which is the same result we found from counting the different scenarios.

The total number of combinations for the outcomes of two independent events is the product of the number of outcomes for each event individually. An **independent** event is one in which the outcome is not changed by the outcome of another event. For example, each flip of a coin is an independent event. Whether the coin lands on heads or tails does not depend on whether it previously landed on heads or tails in the past flips. To understand if an event is not independent, consider a bag of six blue and six red marbles. If we remove one marble from the bag at random, and do not replace it, the next marble that is removed is no longer independent from the previous action. The probability for a blue marble to be removed in the second draw depends on whether the first marble was blue or red.

When combining two or more independent events, the total number of outcomes is simply the product of the probabilities of each individual event. This means, that if we roll 2 dice, we have 36 possible outcomes, and if we roll 3 dice, we have 216 possible outcomes, etc. The probability for each of the outcomes to occur in

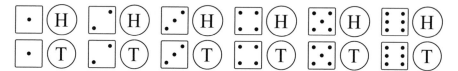

Fig. A.2 All twelve possible outcomes when both rolling a die and flipping a coin

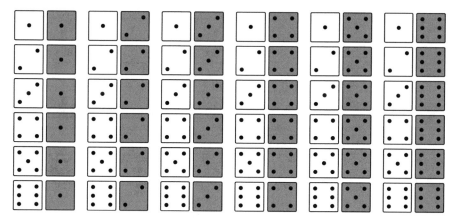

Fig. A.3 All possible outcomes of rolling two dice. One die is colored white and the other red to help distinguish the possible combinations for each die

this case is also the product of the probability for each individual outcome to occur. For example given a red, green, and white die, the probability of rolling a 2 on the red die, a 6 on the green die, and a 1 on the white die is simply the product of the probability for each individual die, or in this case $\frac{1}{216}$.

However, this is not the same when the two events are made to be dependent on each other. Take the situation when we want to find the probability that the sum of two dice adds up to a specific number. All possible outcomes of rolling two dice are shown in Fig. A.3. Here we see that there are at least two ways for the dice to come out so that result in the same number, with the exception of 2 and 12. When we count up the total number of unique sums of the two dice, we get the numbers 2 through 12. The count of the ways each sum can occur is given in Table A.1. In this case, the total number of outcomes is still the product of the number of outcomes for each die, but the probability is not the product of the individual probabilities.

A.2.1 Key Words Review

- **Probability**: The mathematics behind how likely an event is to occur
- **Event**: A group of one or more outcomes that form an observation
- **Trial**: A single outcome for an event in a series of events
- **Sample Space**: The set of all possible outcomes

Table A.1 Possible
outcomes of rolling two dice

Sum of dice	# of combinations	Probability
2	1	$\frac{1}{36}$
3	2	$\frac{2}{36}$
4	3	$\frac{3}{36}$
5	4	$\frac{4}{36}$
6	5	$\frac{5}{36}$
7	6	$\frac{6}{36}$
8	5	$\frac{5}{36}$
9	4	$\frac{4}{36}$
10	3	$\frac{3}{36}$
11	2	$\frac{2}{36}$
12	1	$\frac{1}{36}$

A.2.2   Key Ideas Review

- The number of outcomes for a set of independent events is the product of the number of outcomes for each event.
- The probability for a unique event from independent events is the product of the probability for each independent event
- When considering dependent events, the probability can be calculated by dividing the number of events of interest by the total number of outcomes.

A.2.3   Check Your Understanding

1. Given two dice, what is the probability that when rolled they
   (a) have different numbers?
   (b) sum to a number greater than 8? sum to a number less than 8?
   (c) are both an even number? both an odd number? one even and one odd?
2. Write out all possible outcomes of flipping 5 coins. Calculate the probability for 0, 1, 2, 3, 4, and 5 heads to come out. Plot these probabilities as a histogram.
3. Given a deck of cards, and a die. Calculate the total number of outcomes. Calculate the probabilities for
   (a) Drawing a red card and rolling an even number.
   (b) Drawing a face card and rolling a one.
   (c) Drawing the ace of spades and rolling 4 or higher.

A.3 Histograms

There are many different ways to represent the data collected from a series of events. First, one can represent data as a frequency table. A **frequency table** contains two columns, where the first column is the possible categories that the event can fall into, and the second column is the number of times that event has occurred. It is possible to represent a frequency table as a plot by having the x-axis represent the different categories, and the y-axis representing the frequency of those events. This graphical representation of the frequency table is known as a **histogram**.

To make these concepts more concrete, consider the following example. After flipping five coins, we count the number of heads and find that we have three heads. Now we repeat this trail 29 more times, and at the end we have the following counts for the number of heads for the 30 trials:

$$3, 4, 2, 1, 1, 1, 4, 1, 4, 3,$$
$$2, 3, 2, 2, 3, 2, 2, 1, 3, 3,$$
$$2, 3, 4, 4, 3, 2, 1, 3, 3, 2.$$

We can put this into a frequency table as given in Table A.2.

We can then make a bar plot of the categories, and the frequency of each of those categories. The results for this example is shown in Fig. A.4.

A.4 Mean, Median, and Mode

Given a set of observations of events, it is important to calculate properties of distribution of events. The group of calculations giving a central value are referred to as **measures of central tendency**. These measures, along with other statistical measures of the data, allow us to compare a given dataset to another dataset and draw conclusions. In statistics, there are three common ways of measuring the central tendency. Each measure gives a different piece of information about the distribution.

The first method of measuring central tendency is known as the mean. The **mean** of a set of events is the sum of all the data points representing the events divided by the number of events in the sample. Often, the mean is also called the average. A typical use that you are familiar with is the average grade from a given class.

Table A.2 Frequency table for the number of heads from 30 trails of 5 coin flips

Number of heads	Frequency
0	0
1	6
2	9
3	10
4	5
5	0

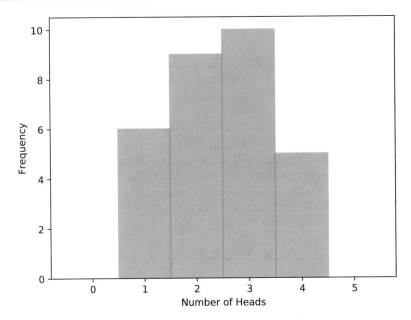

Fig. A.4 Histogram for the number of heads from 30 sets of 5 coin flips

The second method is known as the median. The **median** of a set of events is the middle number when the events are ordered from the smallest value to the largest value. If there happen to be two numbers in the middle, the median is the average of those two values.

The last method is known as the mode. The **mode** is simply the result in a set of events that occurs the most often.

To help make these definitions more clear, let us consider an example. A teacher gave a test to a class of 15 students, and the scores that the students achieved were 78, 100, 81, 89, 91, 88, 93, 99, 85, 75, 95, 94, 84, 81, and 87. First, we will calculate the mean (which is often represented by the symbol μ). Recall, the mean is found by taking the sum of the scores and dividing by the number of scores. The mean for these scores is:

$$\mu = \frac{78+100+81+89+91+88+93+99+85+75+95+94+84+81+87}{15}$$
$$= \frac{1320}{15} \tag{A.1}$$
$$= 88 \ .$$

The mean or average score for this test is 88. Now, to calculate the median, we first sort the data from smallest to largest, and then find the value in the middle. In the example it is given by

$$\underbrace{75, 78, 81, 81, 84, 85, 87,}_{\text{7 numbers}} \overset{\uparrow}{88}, \underbrace{89, 91, 93, 94, 95, 99, 100}_{\text{7 numbers}} \qquad (A.2)$$

$$\text{median}$$

Therefore, the median is also given by the number 88. Finally, the mode is the number that occurs the most often. Table A.3 shows the counts for each of the different test scores, and from that we can see that the mode is the score 81.

Putting together the results for these tests, we find that both the mean score and the median score are 88 and the mode score was 81. These calculations can be extended to arbitrarily large data sets, and the techniques discussed above will still work to calculate the mean, median, and mode. In our example above, the mean and median turn out to be the exact same number. However, this is not always the case. When these two numbers are far away from each other, it could indicate that there is an outlier in the data set. An **outlier** is a point in the data set that drastically differs from the rest of the data. An example of an outlier would be to consider the test scores again. If we were to add in a test score of 5, this would be considered an outlier. Consider the impact this outlier would have on the mean, median, and mode. The mean would now be 82.8 instead of 88, while the median would be 87.5 and the mode would still be 81. From this, we can see that by combining the mean, median, and mode we learn more information about the data than by any one individual measure.

A.4.1   Key Word Review

- **Measures of central tendency**: The group of calculations which provide information about a central value
- **Mean**: The sum of all the data points divided by the total number of data points
- **Median**: The data point in the middle when the data is sorted from smallest to largest
- **Mode**: The data point that occurs the most often in a set of data

A.4.2   Check Your Understanding

For all of the problems below, consider the 30 test scores found in Table A.4.

1. Calculate the mean, median, and mode.

Table A.3 The counts for each test score from the set of 15 tests

Score	75	78	81	84	85	87	88	89	91	93	94	95	99	100
Count	1	1	2	1	1	1	1	1	1	1	1	1	1	1

Test Scores

62	69	82	95	75	60	56	94	81	97
70	72	67	63	68	70	85	72	90	80
61	79	62	88	72	75	86	87	97	72

2. What happens to the mean, median, and mode if 5 new test scores are added with the values of: 6, 12, 20, 14, and 5

A.5   Variance and Standard Deviation

In the previous section, we discussed commonly used measures of central tendency. In this section, we will discuss **measures of dispersion**, which provide information about the spread of data around the central values.

Here, we will cover two of the many measures of dispersion that exist. These two measures are related to each other and are known as the variance and standard deviation. The **variance** is the average squared distance that exists between all the data points and the mean of the data points. The **standard deviation** is the square root of the variance, and gives an impression of the spread of the data around the mean. The smaller the standard deviation, the more clustered the points are around the mean value. The values of the variance and the standard deviation are can never be negative. The variance is typically represented by the symbol σ^2, while the standard deviation is represented by the symbol σ. To calculate the variance and standard deviation we use the following formula

$$\sigma^2 = \frac{\sum (x_i - \mu)^2}{n}, \tag{A.3}$$

where the sum is over all n data points, x_i is the value of each point, and μ is the mean of the data set. As an example, let us look at the test scores given in the previous section: 78, 100, 81, 89, 91, 88, 93, 99, 85, 75, 95, 94, 84, 81, and 87. To find the variance and the standard deviation, we first calculate the mean, which was done in the previous section and is 88. Next we take the difference between each separate score and the mean value, then square each difference. Finally, we sum these squared values and divide them by the number of scores. The steps are shown here as

$$(x_i - \mu)^2 = 100, 144, 49, 1, 9, 0, 25, 121, 9, 169, 49, 36, 16, 49, 1$$

$$\sum (x_i - \mu)^2 = 778$$

$$\sigma^2 = \frac{778}{15} = 51.9 \tag{A.4}$$

$$\sigma = \sqrt{\sigma^2} = 7.2$$

The standard deviation is 7.2, and this means that most of the scores are within 7.2 points of 88.

A.5.1   Key Word Review

- **Measures of dispersion**: Calculations which provide information on the spread of the data
- **Variance**: The average of the squared distance to the mean of a dataset
- **Standard Deviation**: The square root of the variance

A.5.2   Check Your Understanding

For all the problems below, again consider the 30 test scores found in Table A.4.

1. Calculate the variance and the standard deviation
2. What happens to the variance and the standard deviation if 5 new test scores are added with the values of: 6, 12, 20, 14, and 5.

Linear Algebra

B

B.1 Introduction

Linear algebra is the field of mathematics underlying the topics of vectors and matrices. It is the backbone of mathematics used for solving systems of linear equations. You may be familiar with static physics problems, whose solution utilize linear algebra. An example of a set of linear equations is

$$x_1 + 4x_2 + 3x_3 = -7$$
$$5x_1 + 2x_2 = 11 \tag{B.1}$$
$$9x_2 - 8x_3 = 0$$

This appendix supplies a basic introduction to some key aspects of the field of linear algebra. We will cover the topics of scalars, vectors, and matrices.

B.2 Scalars, Vectors, and Matrices

B.2.1 Scalars

A **scalar** is a number without a direction associated to it. Some common examples of scalars are speed and distance. Both speed and distance are quantified by a single number, but give no information about the direction one is traveling. In this book, the scalars that are used are either real numbers or complex numbers. Let us quickly review complex numbers since they are not as common as real numbers in everyday life.

A complex number consists of a real and an imaginary component. An **imaginary** number is one that is some multiple of $i = \sqrt{-1}$, i.e. $4i$, $1/10i$, and πi are all examples of imaginary numbers. Complex numbers are represented using the form

© The Author(s) 2021
C. Hughes et al., *Quantum Computing for the Quantum Curious*,
https://doi.org/10.1007/978-3-030-61601-4

$a + bi$, with both a and b as real numbers. Addition and subtraction of complex numbers is straightforward, we simply add or subtract the real components from each other and add or subtract the imaginary components from each other. Some examples are:

$$(3 + 4i) + (5 + 3i) = 8 + 7i$$
$$(3 + 4i) - (1 + 3i) = 2 + i$$
$$(3 + i) + (1 - 3i) = 4 - 2i$$
$$(9 - 6i) - (4 + i) = 5 - 7i$$

On the other hand, multiplication and division are slightly more complicated. When performing multiplication and division, it is helpful to know the powers of i:

$$i^2 = -1, \quad i^3 = -i, \quad i^4 = 1, \quad i^5 = i, \quad \text{etc.} \tag{B.2}$$

Multiplication of complex numbers is performed by multiplying out the terms and then adding them together, i.e. given the complex numbers $x = a + bi$ and $y = c + di$, then:

$$x \times y = (a + bi) \times (c + di) = (ac + adi + bci + bdi^2) = (ac - bd) + (ad + bc)i. \tag{B.3}$$

Before covering division, let us introduce the idea of a complex conjugate. A **complex conjugate** is the complex number where the real component is the same, but the imaginary component has the opposite sign. The complex conjugate is denoted with an $*$ symbol. For example, the complex conjugate of $a + bi$ is $a - bi$. When multiplying a complex number by its complex conjugate we get a real number, and refer to it as the **magnitude squared** of a complex number and is represented by $|x|^2$, i.e. $|a + bi|^2 = (a + bi) \times (a + bi)^* = a^2 + b^2$.

In order to divide two complex numbers, first you multiply the numerator and denominator by the complex conjugate of the denominator and then perform multiplication of complex numbers, e.g.,

$$\frac{a + bi}{c + di} = \frac{a + bi}{c + di} \times \frac{c - di}{c - di} = \frac{(a + bi) \times (c - di)}{c^2 + d^2} = \frac{(ac + bd) + (bc - ad)i}{c^2 + d^2}. \tag{B.4}$$

B.2.2   Vectors

A **vector** is a mathematical object that is defined to have both size and direction. One can imagine a vector as a directed line segment as seen in Fig. B.1. The size or **magnitude** of a vector is the length of the line segment. The **tail** of a vector is the starting point, and the **head** of a vector is the ending point. The **direction** of a

Fig. B.1 Geometric depiction of a vector

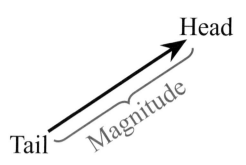

Fig. B.2 Depiction of a vector on a coordinate system in 2 dimensions

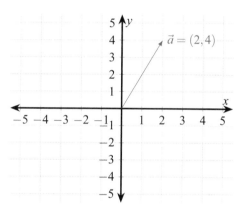

vector is from the tail to the head. Two vectors are considered equal to each other when they both have the same magnitude and the same direction.

Some common vectors that you may be familiar with are force, velocity, and acceleration. For example, a car's velocity is the speed the car is going combined with the direction the car is driving in. Typically, vectors are denoted by either $\vec{a}$ or **a**, and the magnitude of a vector is denoted by $|\vec{a}|$. The magnitude is a scalar value (see Sect. B.2.1 for details). Additionally, we can describe a vector by imagining a coordinate system, where the origin is located at the tail of the vector (in two dimensions this would correspond to the point $(0, 0)$), and the head of the vector would be its coordinate (e.g., (x, y) in two dimensions). For example, given $\vec{a}$ as a two dimensional vector as shown in Fig. B.2, we can define the vector as $\vec{a} = (2, 4)$.

The magnitude of a vector can be calculated by using the distance formula between two points. Recall that the distance formula between the origin of a coordinate system in two dimensions is given by

$$d = \sqrt{x^2 + y^2}. \tag{B.5}$$

The formula for the distance from the origin of a vector to its endpoint can be generalized to higher dimensions in a straightforward manner. For example, given a ten dimensional vector $\vec{a} = (x_1, x_2, x_3, x_4, x_5, x_6, x_7, x_8, x_9, x_{10})$, the magnitude is given by

$$|\vec{a}| = \sqrt{x_1^2 + x_2^2 + x_3^2 + x_4^2 + x_5^2 + x_6^2 + x_7^2 + x_8^2 + x_9^2 + x_{10}^2}. \qquad (B.6)$$

In this book, we mainly focus on qubits which we represent as two-dimensional vectors. Qubits are also unit normalized, meaning (x, y) satisfies $x^2 + y^2 = 1$. If a vector (a, b) is not unit normalized, then the vector $(a/(a^2 + b^2), b/(a^2 + b^2))$ is unit normalized.

Operations with Vectors

Consider what we can do with one vector and one scalar, or two vectors. First, let us consider the addition of two vectors. If we have the vectors $\vec{a}$ and $\vec{b}$, we can calculate the sum $\vec{c} = \vec{a} + \vec{b}$ by adding each component independently. For example, given the two dimensional vectors $\vec{a} = (x_1, y_1)$, and $\vec{b} = (x_2, y_2)$, the sum of the vectors is $\vec{c} = \vec{a} + \vec{b} = (x_1 + x_2, y_1 + y_2)$. This can be represented pictorially and is shown in Fig. B.3.

There are two important properties of vector addition:

- Vector addition is **commutative**, or in words the order does not matter (i.e. $\vec{a} + \vec{b} = \vec{b} + \vec{a}$). This is often referred to as the parallelogram law, and can be pictorially shown to be true in Fig. B.4.
- Vector addition is **associative**, or in words when summing three vectors it does not matter which pair is added together first (i.e. $(\vec{a} + \vec{b}) + \vec{c} = \vec{a} + (\vec{b} + \vec{c})$).

Fig. B.3 Demonstration of vector addition

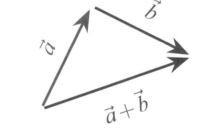

Fig. B.4 The parallelogram law

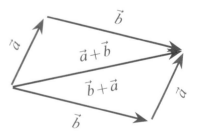

Subtraction of two vectors works in the same way as addition, but instead of adding the components, we subtract them. This can also be thought of taking a vector $\vec{a}$, and flipping the direction of the vector. This will change $\vec{a} = (x, y)$ to $-\vec{a} = (-x, -y)$. All of the rules for addition described above also apply to subtraction.

If we have a vector and a scalar, the only thing we can do is to multiply the two together. When multiplying a vector and a scalar, we simply multiply every component of the vector by the scalar. For example, given the three dimensional vector $\vec{a} = (4, 6, 2)$ and the scalar $s = 5$, then $\vec{b} = s \times \vec{a} = (20, 30, 10)$. It is important to notice here that the result $\vec{b}$ is also a vector. The direction of the vector $\vec{b}$ is unchanged if $s > 0$, but is in the opposite direction if $s < 0$. The magnitude of the vector $\vec{b}$ is scaled by the value of the scalar, and thus it is stretched if $|s| > 1$ and compressed if $|s| < 1$.

Finally, we have the multiplication of two vectors. There are actually two different ways that we can multiply vectors together. The first way is known as the **dot product**. The dot product takes two vectors, and multiplies them together to produce a scalar. We denote the dot product of two vectors by $\vec{a} \cdot \vec{b}$. The dot product is calculated by taking the product of all the components of the vector, and summing them up. In other words, given the vector $\vec{a} = (x_a, y_a)$ and the vector $\vec{b} = (x_b, y_b)$, the dot product is given as: $\vec{a} \cdot \vec{b} = x_a \times x_b + y_a \times y_b$. While this example is in two dimensions, the dot product can be extended to an arbitrary number of dimensions.

We can also think about the dot product geometrically. Let us take two vectors $\vec{a}$ and $\vec{b}$, with the magnitude of $\vec{b}$ larger than that of $\vec{a}$. We can then relate the dot product to the projection of $\vec{a}$ onto the direction of $\vec{b}$, or what portion of $\vec{a}$ is pointing in the direction of $\vec{b}$. The projection can be thought of as imagining what the shadow of $\vec{a}$ would look like when cast by a light source perpendicular to $\vec{b}$ (see Fig. B.5). In other words, the dot product can be represented by multiplying the magnitudes of the two vectors, and taking the cosine of the angle between them ($\vec{a} \cdot \vec{b} = |\vec{a}||\vec{b}| \cos \theta$). Below are some properties that are related to dot products, where $\vec{a}, \vec{b}$, and $\vec{c}$ are vectors, and s is a scalar:

- $\vec{a} \cdot \vec{b} = \vec{b} \cdot \vec{a}$
- $(\vec{a} + \vec{b}) \cdot \vec{c} = \vec{a} \cdot \vec{c} + \vec{b} \cdot \vec{c}$
- $(s\vec{a}) \cdot \vec{b} = s(\vec{a} \cdot \vec{b})$

The other method of multiplying two vectors is known as the cross product. The cross product is written as $\vec{a} \times \vec{b}$. Note that the notation for the cross product looks the same as multiplication. However, multiplication operates on scalars while the cross products acts on vectors. The **cross product** takes two vectors, and produces a third vector that is perpendicular to the other two. From this definition, it can easily be seen that this will not work for two dimensional vectors, so we will focus on three dimensions for this discussion. The magnitude of the resulting vector from the cross product is equal to the area of the parallelogram formed by the vectors (i.e. $|\vec{a} \times \vec{b}| = |\vec{a}||\vec{b}| \sin \theta$ and as shown in Fig. B.6).

Fig. B.5 Geometric
representation of the dot
product

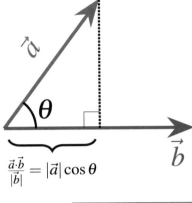

$$\frac{\vec{a} \cdot \vec{b}}{|\vec{b}|} = |\vec{a}| \cos \theta$$

Fig. B.6 Relationship
between the magnitude of the
cross product and area of
parallelogram

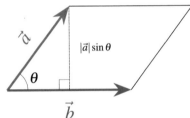

The direction of the resulting vector from a cross product can be determined by the **right-hand rule**. To use the right-hand rule given the cross product $\vec{c} = \vec{a} \times \vec{b}$, first point the fingers of the right hand in the direction of $\vec{a}$, and curl them towards $\vec{b}$. The direction the thumb points is the direction of $\vec{c}$. Using the right-hand rule, we can make an observation that $\vec{a} \times \vec{b} = -\vec{b} \times \vec{a}$. From this, we can also see that $\vec{a} \times \vec{a} = -\vec{a} \times \vec{a}$, which is only possible if the result is the 0 vector (i.e. $\vec{c} = (0, 0, 0)$). Given the vectors $\vec{a} = (x_1, y_1, z_1)$ and $\vec{b} = (x_2, y_2, z_2)$, the cross product can be calculated using:

$$\vec{c} = \vec{a} \times \vec{b} = (y_1 z_2 - y_2 z_1, x_2 z_1 - x_1 z_2, x_1 y_2 - x_2 y_1). \tag{B.7}$$

B.2.3   Matrices

A **matrix** (plural matrices) is a rectangular array of numbers, typically written in rows and columns with round brackets around them. We denote the size of the matrix by first listing the number of rows, followed by the number of columns. For example,

$$M = \begin{pmatrix} 3 & 7 \\ 1 & 4 \\ 9 & 5 \end{pmatrix}, \tag{B.8}$$

is a 3×2 matrix. Three dimensional vectors (discussed in the previous section) can be thought of as either a 1×3 or a 3×1 matrix. Often the number of rows and columns of the matrix are referred to as the **shape** of the matrix. We say a matrix is a **square matrix** if the number of rows is equal to the number of columns. We can specify an element of a matrix by denoting its row and column position. For example, taking the matrix M given in Eq. (B.8), the element $M_{1,2} = 7$ and the element $M_{3,1} = 9$. This representation of the matrix is often referred to as index notation, and will be used below to make the equations easier to understand.

Addition and Subtraction of Matrices
Addition and subtraction of matrices is straightforward. The only requirement on addition and subtraction of two matrices is that they must have the shape (number of rows and number of columns). When adding or subtracting matrices of the same shape, one performs each operation between the same element location. For example, given the 3×3 matrices

$$A = \begin{pmatrix} 9 & 5 & -6 \\ 1 & -10 & -8 \\ 1 & 5 & -3 \end{pmatrix}, \quad B = \begin{pmatrix} -4 & 7 & -5 \\ -9 & -9 & 5 \\ 9 & -1 & 8 \end{pmatrix}, \tag{B.9}$$

then

$$A + B = \begin{pmatrix} 9+(-4) & 5+7 & -6+(-5) \\ 1+(-9) & -10+(-9) & -8+5 \\ 1+9 & 5+(-1) & -3+8 \end{pmatrix} = \begin{pmatrix} 5 & 12 & -11 \\ -8 & -19 & -3 \\ 10 & 4 & 5 \end{pmatrix}, \tag{B.10}$$

and

$$A - B = \begin{pmatrix} 9-(-4) & 5-7 & -6-(-5) \\ 1-(-9) & -10-(-9) & -8-5 \\ 1-9 & 5-(-1) & -3-8 \end{pmatrix} = \begin{pmatrix} 13 & -2 & -1 \\ 10 & -1 & -13 \\ -8 & 6 & -11 \end{pmatrix}. \tag{B.11}$$

Multiplication with Matrices
When dealing with matrices, there are three types of multiplication that we will discuss. These are the multiplication of scalars and matrices, the multiplication of vectors and matrices, and the multiplication of matrices and matrices.

The result of multiplication by a scalar is just an element-wise multiplication by the scalar, e.g.,

$$A = \begin{pmatrix} 6 & 10 & 0 \\ 5 & 7 & 1 \\ 5 & 6 & 1 \end{pmatrix}, \quad 5A = \begin{pmatrix} 5 \times 6 & 5 \times 10 & 5 \times 0 \\ 5 \times 5 & 5 \times 7 & 5 \times 1 \\ 5 \times 5 & 5 \times 6 & 5 \times 1 \end{pmatrix} = \begin{pmatrix} 30 & 50 & 0 \\ 25 & 35 & 5 \\ 25 & 30 & 5 \end{pmatrix}. \tag{B.12}$$

If we think about vectors as $1 \times N$ or $N \times 1$ shaped matrices, then the product of a matrix and a vector can be thought of as the product of two matrices. It is easiest to build up to the definition of matrix-matrix multiplication with the use of examples. First, the dot product discussed in Sect. B.2.2 can be thought about as multiplying a $1 \times N$ matrix by a $N \times 1$ matrix. For example, with

$$A = \begin{pmatrix} 2 \\ 9 \end{pmatrix}, \text{ and } B = \begin{pmatrix} 3 & 5 \end{pmatrix}, \tag{B.13}$$

then

$$B \times A = (3 \times 2) + (5 \times 9) = (51). \tag{B.14}$$

Note that we started with matrix B (1×2) and matrix multiplied it with A (2×1) to find 1×1 matrix. This is the same result as the dot product between two vectors $\vec{a} = (2, 9)$ and $\vec{b} = (3, 5)$. We can extend this by breaking down the matrix into a set of $1 \times N$ and $N \times 1$ matrices. It is important to note that the number of columns of the first matrix must be equal to the number of rows of the second matrix for matrix multiplication to make sense. Given a $m \times n$ matrix A and a $n \times p$ matrix B, then the product $A \times B = C$ is a $m \times p$ matrix. The (i, j) element of the matrix C can be calculated as

$$c_{i,j} = \sum_{k=1}^{n} a_{i,k} \times b_{k,j}, \tag{B.15}$$

where the $\sum_{k=1}^{n}$ symbol means we sum all indices k from 1 to n. As an example, let us calculate $A \times B = C$ and $B \times A = D$ for the 2×2 matrices given by

$$A = \begin{pmatrix} 3 & 2 \\ 5 & 6 \end{pmatrix}, \quad B = \begin{pmatrix} 2 & 6 \\ 8 & 7 \end{pmatrix}. \tag{B.16}$$

First, let us calculate the matrix C by treating the ith row of A and the jth column of B as if they were vectors as done above:

$$c_{1,1} = \begin{pmatrix} 3 & 2 \end{pmatrix} \begin{pmatrix} 2 \\ 8 \end{pmatrix} = (3 \times 2 + 2 \times 8) = 22$$

$$c_{1,2} = \begin{pmatrix} 3 & 2 \end{pmatrix} \begin{pmatrix} 6 \\ 7 \end{pmatrix} = (3 \times 6 + 2 \times 7) = 32$$

$$c_{2,1} = \begin{pmatrix} 5 & 6 \end{pmatrix} \begin{pmatrix} 2 \\ 8 \end{pmatrix} = (5 \times 2 + 6 \times 8) = 58$$

$$c_{2,2} = \begin{pmatrix} 5 & 6 \end{pmatrix} \begin{pmatrix} 6 \\ 7 \end{pmatrix} = (5 \times 6 + 6 \times 7) = 72$$

Taken together,

$$C = \begin{pmatrix} 22 & 32 \\ 58 & 72 \end{pmatrix}.$$ (B.17)

Now, by following similar steps we can calculate $B \times A$ by taking the ith row of B and the jth column of A to produce

$$d_{1,1} = \begin{pmatrix} 2 & 6 \end{pmatrix} \begin{pmatrix} 3 \\ 5 \end{pmatrix} = (2 \times 3 + 6 \times 5) = 36$$

$$d_{1,2} = \begin{pmatrix} 2 & 6 \end{pmatrix} \begin{pmatrix} 2 \\ 6 \end{pmatrix} = (2 \times 2 + 6 \times 6) = 40$$

$$d_{2,1} = \begin{pmatrix} 8 & 7 \end{pmatrix} \begin{pmatrix} 3 \\ 5 \end{pmatrix} = (8 \times 3 + 7 \times 5) = 59$$

$$d_{2,2} = \begin{pmatrix} 8 & 7 \end{pmatrix} \begin{pmatrix} 2 \\ 6 \end{pmatrix} = (8 \times 2 + 7 \times 6) = 58$$

Taken together, the result is

$$D = \begin{pmatrix} 36 & 40 \\ 59 & 58 \end{pmatrix}.$$ (B.18)

From this, we can see that the result of $A \times B = C \neq D = B \times A$, or in other words, that matrix multiplication is not **commutative**.

Transposition

There also exists operations on a single matrix. For this textbook, the only one that is important to cover is matrix transposition. The operation of the matrix **transpose** simply turns the rows into columns and the columns into rows. Mathematically, the transpose is represented by A^T and can be represented using the index notation for a matrix A as

$$A^T = (a_{i,j})^T = a_{j,i}.$$ (B.19)

For example, given the matrix

$$A = \begin{pmatrix} 9 & 7 & 2 & 6 \\ 9 & 3 & 4 & 4 \\ 2 & 8 & 8 & 4 \end{pmatrix},$$

the transpose of A is given as:

$$A^T = \begin{pmatrix} 9 & 9 & 2 \\ 7 & 3 & 8 \\ 2 & 4 & 8 \\ 6 & 4 & 4 \end{pmatrix}.$$

Note that this changes a $m \times n$ matrix into a $n \times m$ matrix.

B.3 Key Word Review

- **Scalar**: a physical quantity that is completely described by its magnitude.
- **Vector**: a quantity which has a direction as well as a magnitude.
- **Matrix**: a collection of numbers arranged into a fixed number of rows and columns.

B.4 Check Your Understanding

1. Find the dot product of the following two vectors:

$$\begin{pmatrix} 1 \\ 4 \\ 3 \end{pmatrix} \quad \begin{pmatrix} -3 \\ 2 \\ 5 \end{pmatrix}$$

2. Find the cross product of $v_1 \times v_2$:

$$v_1 = \begin{pmatrix} 1 \\ 10 \\ 7 \end{pmatrix} \quad v_2 = \begin{pmatrix} 2 \\ 2 \\ 8 \end{pmatrix}$$

3. Perform the matrix operation $\sigma_1 \times \sigma_2$:

$$\sigma_1 = \begin{pmatrix} 0 & 1 \\ 1 & 0 \end{pmatrix} \quad \sigma_2 = \begin{pmatrix} 0 & -i \\ i & 0 \end{pmatrix}$$

4. Transpose the following matrix:

$$\begin{pmatrix} 1 & -2 & 4 \\ 5 & 3 & 8 \\ 1 & 9 & 0 \end{pmatrix}$$

Answers to Odds

Chapter 1 Solutions

1. (a) quantized to the charge of the electron: $e = 1.6 \times 10^{-19}$ C.
   (b) time is continuous
   (c) space is continuous
   (d) quantized to $0.01
   (e) continuous because the frequency of light (which causes color) is continuous
3. No. If we showed 100 copies of the picture to Student A, they would always see blue/black. In a 50/50 quantum superposition, they would each see around 50 pictures as blue/black and the rest as white/gold. The two states must be an intrinsic property of the dress rather than something that depends on the observer.

Chapter 2 Solutions

1. With the 8-bit representation, this would require eight coins arranged as TTTTT-THT.
3. (a) 9/10 or 90%. The probability is the square of the amplitude.
   (b) No, $1/9 + 4/9 \neq 1$. Normalization means the total probabilities add up to 1.
   (c) 75 coins.
   (d) Measurement collapses the superposition onto either $|H\rangle$ or $|T\rangle$.
5. (a) All of them are possible as they all can produce $|0\rangle$ after measurement.
   (b) After measurement, the state in the question is in $|0\rangle$, since the superposition collapsed. Since it is in the $|0\rangle$ state after measurement, if you try to measure the same state again you will always measure $|0\rangle$. No new information is provided about the state after the collapse to $|0\rangle$.
   (c) If $|0\rangle$ is measured from the unknown state, and a second identical state is prepared and is measured in the $|1\rangle$ state, then you know the unknown state contains some nonzero superposition of both $|0\rangle$ and $|1\rangle$, e.g., it is $|\psi\rangle = \alpha|0\rangle + \beta|1\rangle$ with $\alpha \neq 0$ and $\beta \neq 0$. So you can rule out $|\psi\rangle = |0\rangle$ as the initial state, but all of the other three states given in (a) are still possible.

© The Author(s) 2021
C. Hughes et al., *Quantum Computing for the Quantum Curious*,
https://doi.org/10.1007/978-3-030-61601-4

Measurements of many more particles are needed to determine the numerical values of α and β in order to find the exact state.

$\begin{pmatrix} 0 \\ 1 \end{pmatrix}$, which is the $|1\rangle$ state.

7. $\begin{pmatrix} 1 \\ 0 \end{pmatrix}$, which is the $|0\rangle$ state.

9.

$$Y^\dagger = \begin{pmatrix} 0 & -i \\ i & 0 \end{pmatrix}. \tag{1}$$

11. Take $|\psi\rangle = |0\rangle$. The probability of measuring $|\psi\rangle$ in the $|0\rangle$ state is 100%, and the probability of measuring $|\psi\rangle$ in the $|1\rangle$ state is 0%. Here the probabilities add up to 100%. Now take any random non-unitary matrix, for example,

$$N = \begin{pmatrix} 0 & 1 \\ 0 & 0 \end{pmatrix}. \tag{2}$$

Applying this non-unitary matrix to $|\psi\rangle$ gives the state

$$N|\psi\rangle = \begin{pmatrix} 0 & 1 \\ 0 & 0 \end{pmatrix} \begin{pmatrix} 1 \\ 0 \end{pmatrix} = \begin{pmatrix} 0 \\ 0 \end{pmatrix} = 0|0\rangle + 0|1\rangle. \tag{3}$$

This new state after non-unitary evolution has 0% probability of being in the $|0\rangle$ state, and 0% probability of being in the $|1\rangle$ state. The total probability does **not** add up to 100%.

Chapter 3 Solutions

1. All parts of the statement are false. The particle is in both states the entire time before measurement and has a 50% chance of being measured as 0 or 1.
3. a. The coincidence counts are low when the detectors are an equal distance from the beam splitter. This points to light behaving like a particle entering only one detector at a time.
   b. Photons from different 0.4 μs bursts can arrive at the detectors simultaneously.
5. A 30/70 superposition state would take the form:

$$\sqrt{\frac{3}{10}}|0\rangle + \sqrt{\frac{7}{10}}|1\rangle. \tag{4}$$

The desired beam splitter matrix M should perform the operation:

$$M \begin{pmatrix} 0 \\ 1 \end{pmatrix} = \begin{pmatrix} \sqrt{\frac{3}{10}} \\ \sqrt{\frac{7}{10}} \end{pmatrix} \implies M = \begin{pmatrix} 1 & \sqrt{\frac{3}{10}} \\ 1 & \sqrt{\frac{7}{10}} \end{pmatrix} \tag{5}$$

would give the correct probabilities, but it is not unitary. And all quantum matrices must be unitary. Using the 50/50 beamsplitter as a reference, the unitary 30/70 beam splitter matrix is

$$M = \begin{pmatrix} \sqrt{\frac{7}{10}} & \sqrt{\frac{3}{10}} \\ -\sqrt{\frac{3}{10}} & \sqrt{\frac{7}{10}} \end{pmatrix}. \tag{6}$$

Chapter 4 Solutions

1.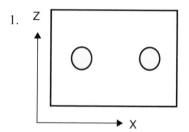

3. (a) $\cos(\pi/6)|0\rangle + \sin(\pi/6)|1\rangle$
   (b) $\alpha^2 = \cos^2(\pi/6) = 0.75$
   (c) $\sin(\pi/6 + \pi/4)|+\rangle + \cos(\pi/6 + \pi/4)|-\rangle$
   (d) $\cos^2(\pi/6 + \pi/4) \approx 0.067$
5. 100% $+z$
7. 50% up, 50% down
9. 50% $+x$, 50% $-x$

Chapter 5 Solutions

1. Their basis will match approximately half of the time, so their key will be approximately 500,000 bits long. The number of matching bits will not be exactly 500,000 due to the probabilistic nature of measuring up and down randomly in a finite sample size, i.e., think of flipping a coin. However, the margin of error at the 95% confidence interval is $\approx 0.03\%$, meaning that with one million measurements of qubits in a 50-50 superposition, we would expect the margin of error to be 300 bits.
3. $(1/2)^{20} \approx 10^{-6}$.
5. If Alice and Bob both use the z-basis, the different cases are:
   (a) Alice sends $+z$, Eve measures in z, Bob measures $+z$. ✓

(b) Alice sends -z, Eve measures in z, Bob measures -z. ✓
(c) Alice sends +z, Eve measures in x, Bob measures +z (will happen with 50% probability). ✓
(d) Alice sends +z, Eve measures in x, Bob measures -z (will happen with 50% probability).
(e) Alice sends -z, Eve measures in x, Bob measures +z (will happen with 50% probability).
(f) Alice sends -z, Eve measures in x, Bob measures -z (will happen with 50% probability). ✓

Therefore there is a 4/6 probability that Eve has not been detected.

7. Copy the state of each electron, passing the originals along to Bob. Once the correct basis is revealed publicly, pass your cloned electrons through SGAs oriented in the correct revealed basis and get the key.

Chapter 6 Solutions

1. (a)

$$X|0\rangle = \begin{pmatrix} 0 & 1 \\ 1 & 0 \end{pmatrix} \begin{pmatrix} 1 \\ 0 \end{pmatrix} = \begin{pmatrix} 0 \\ 1 \end{pmatrix} = |1\rangle. \tag{7}$$

(b)

$$X|\psi\rangle = \begin{pmatrix} 0 & 1 \\ 1 & 0 \end{pmatrix} \begin{pmatrix} \alpha \\ \beta \end{pmatrix} = \begin{pmatrix} \beta \\ \alpha \end{pmatrix} = \beta|0\rangle + \alpha|1\rangle. \tag{8}$$

3. B; the superposition has collapsed into a $|1\rangle$ state after measurement.

5. Sending spin up electrons into a horizontal SGA is identical to applying a Hadamard gate to a $|0\rangle$ qubit.

7. (a)

$$H|1\rangle = \frac{1}{\sqrt{2}} \begin{pmatrix} 1 & 1 \\ 1 & -1 \end{pmatrix} \begin{pmatrix} 0 \\ 1 \end{pmatrix} = \frac{1}{\sqrt{2}} \begin{pmatrix} 1 \\ -1 \end{pmatrix} = |-\rangle. \tag{9}$$

(b)

$$HH|1\rangle = \frac{1}{2} \begin{pmatrix} 1 & 1 \\ 1 & -1 \end{pmatrix} \begin{pmatrix} 1 \\ -1 \end{pmatrix} = \begin{pmatrix} 0 \\ 1 \end{pmatrix} = |1\rangle. \tag{10}$$

(c)

$$HH|\psi\rangle = \frac{1}{2}\begin{pmatrix} 1 & 1 \\ 1 & -1 \end{pmatrix}\begin{pmatrix} 1 & 1 \\ 1 & -1 \end{pmatrix}\begin{pmatrix} \alpha \\ \beta \end{pmatrix} = \begin{pmatrix} \alpha \\ \beta \end{pmatrix}. \tag{11}$$

9.

$$Z|+\rangle = \frac{1}{\sqrt{2}}\begin{pmatrix} 1 & 0 \\ 0 & -1 \end{pmatrix}\begin{pmatrix} 1 \\ 1 \end{pmatrix} = \frac{1}{\sqrt{2}}\begin{pmatrix} 1 \\ -1 \end{pmatrix} = |-\rangle. \tag{12}$$

11. (a) The Z gate does not affect the $|0\rangle$ state.
   (b) The sign on the $|1\rangle$ state is changed, but this does not affect probabilities and so cannot be seen in the histogram.
   (c) The $|+\rangle$ is changed to a $|-\rangle$, which shows up as 50% $|0\rangle$ and $|1\rangle$.
   (d) The $|-\rangle$ is changed to a $|+\rangle$, which shows up as 50% $|0\rangle$ and $|1\rangle$.
13.

$$H^\dagger H = \frac{1}{2}\begin{pmatrix} 1 & 1 \\ 1 & -1 \end{pmatrix}\begin{pmatrix} 1 & 1 \\ 1 & -1 \end{pmatrix} = \frac{1}{2}\begin{pmatrix} 2 & 0 \\ 0 & 2 \end{pmatrix} = \begin{pmatrix} 1 & 0 \\ 0 & 1 \end{pmatrix}. \tag{13}$$

Performing the Hadamard operation twice is the same as multiplying by the identity matrix. Thus, the qubit is unchanged.

Chapter 7 Solutions

1. (a) The probability of measuring $|00\rangle$ is

$$\text{Prob}(|00\rangle) = \left(\frac{1}{\sqrt{2}}\right)^2. \tag{14}$$

We get this by taking the coefficient of the $|00\rangle$ term and then squaring it.
   (b) The probability of measuring the first qubit as 1, Prob(first qubit $|1\rangle$)), is the sum of all outcomes which have the first qubit in the $|1\rangle$ state. In this example, this is Prob(first qubit $|1\rangle$)) = Prob($|10\rangle$) + Prob($|11\rangle$)), which is equal to

$$\left(\frac{1}{2}\right)^2 + \left(-\frac{1}{2}\right)^2 = \frac{1}{2}. \tag{15}$$

(c) The probability of measuring the second qubit as 0, Prob(second qubit $|0\rangle$), is the sum of all outcomes which have the second qubit in the $|0\rangle$ state. In this example, this is Prob(second qubit $|0\rangle$) = Prob($|00\rangle$) + Prob($|10\rangle$), which is equal to

$$\frac{1}{2} + \frac{1}{4} = \frac{3}{4}. \tag{16}$$

(d) After measuring the first qubit as 0, then we know that the only part of $|\psi\rangle$ that has the first qubit as 0 is $(1/\sqrt{2})|00\rangle$. However, we need to renormalize the state to make sure it has a probability of one. So the new state of the system after the measuring the first qubit as 0 is $|\psi'\rangle = |00\rangle$.

(e) After measuring the first qubit as 1, then we know that the only parts of $|\psi\rangle$ that have the first qubit as 1 are $(1/2)|10\rangle - (1/2)|11\rangle$. However, we need to renormalize the state to make sure it has a probability of one. So the new state of the system after the measuring the first qubit as 1 is $|\psi'\rangle = \frac{1}{\sqrt{2}}|10\rangle - \frac{1}{\sqrt{2}}|11\rangle$.

3. Not entangled. Knowing that the first qubit is 0 does not narrow down whether the second qubit is 0 or 1. Qubit 1 is $|0\rangle$ and Qubit 2 is $\frac{1}{\sqrt{2}}(|0\rangle + |1\rangle)$. The combination of the two qubits to form a multi-qubit state gives the state in the question.

5. (a) $|00\rangle$
   (b) $|11\rangle$
   (c) $|01\rangle$
   (d) $\frac{1}{\sqrt{2}}|11\rangle + \frac{1}{\sqrt{2}}|10\rangle$
   (e) $\frac{1}{\sqrt{2}}|00\rangle + \frac{1}{2}|10\rangle - \frac{1}{2}|01\rangle$

7. Note which qubit is the control qubit in IBM convention.
   (a) $|01\rangle$
   (b) $|10\rangle$
   (c) $|11\rangle$
   (d) $|01\rangle$

9. (a) The states change from the start to the end after every gate as: $|00\rangle \rightarrow |01\rangle \rightarrow |11\rangle \rightarrow |10\rangle$.
   (b) $|00\rangle \rightarrow |0\rangle(\frac{1}{\sqrt{2}}|0\rangle + \frac{1}{\sqrt{2}}|1\rangle) \rightarrow \frac{1}{\sqrt{2}}(|00\rangle + |11\rangle) \rightarrow \frac{1}{\sqrt{2}}|0\rangle(\frac{1}{\sqrt{2}}|0\rangle + \frac{1}{\sqrt{2}}|1\rangle)$
   $+ \frac{1}{\sqrt{2}}|1\rangle(\frac{1}{\sqrt{2}}|0\rangle - \frac{1}{\sqrt{2}}|1\rangle) = \frac{1}{2}(|00\rangle + |01\rangle + |10\rangle - |11\rangle)$.

11. No. Alice measures a random value. This automatically changes Bob's state. However, Alice would need to send a classical message to Bob to find out which state Bob measured; therefore information is still bounded by the classical speed of light. This scenario is consistent with Bell's theorem.

Chapter 8 Solutions

1. No, the qubits stay in place. Teleportation only changes the state of an existing qubit.
3. Once it is measured after the Hadamard gate.
5. The X gate flips the $|0\rangle$ into $|1\rangle$ and $|1\rangle$ into $|0\rangle$.
7. The X gate changes the state into $a|0\rangle - b|1\rangle$ and the Z gate flips the sign on the $|1\rangle$ to produce the desired state.

Chapter 9 Solutions

1. (a) Both the classical and quantum computer can represent $2^8 = 256$ classical pieces of information.
   (b) Both the classical and quantum computer can represent $2^8 = 256$ classical pieces of information.
   (c) The quantum computer can create a superposition of up to 256 possibilities and do a computation on all of them. However, a measurement will be only one classical value.
3. (a) Detector 1 only
   (b) Detector 2 only
   (c) Only one photon is needed.
5. (a) The most naive classical algorithm is one where you evaluate every function value and make a list of the results. In this case there are $2^3 = 8$ different function values that would need to be evaluated separately. If each element in the list is identical then the function is constant.
   (b) Since there are results other than $|000\rangle$ and $|111\rangle$, the function is not constant (Fig. 1).

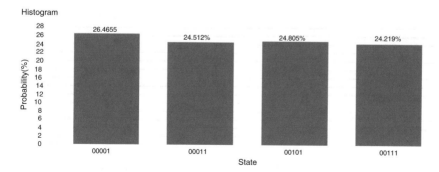

Fig. 1 Reprint courtesy of International Business Machines Corporation, © International Business Machines Corporation.

Chapter 10 Solutions

10.1

1. Alice always measures up; Bob always measures down.
3. (a) Yes.
   (b) Yes. Every time Alice measures up, Bob measures down and vice versa.
5. They each see up and down 50% of the time
7.

$$|\uparrow\downarrow\rangle=\frac{1}{\sqrt{2}}\left(|+\rangle+|-\rangle\right)\times\frac{1}{\sqrt{2}}\left(|+\rangle-|-\rangle\right)=\frac{1}{2}|++\rangle-\frac{1}{2}|+-\rangle+\frac{1}{2}|-+\rangle-\frac{1}{2}|--\rangle.$$

All four possible states are observed. The middle two terms do not cancel out because they are different states: Alice measures $+$ and Bob measures $-$, or Alice measures $-$ and Bob measures $+$.

9. They cannot tell them apart in the z-basis, but they could measure the particles in the x-basis. If Alice and Bob always get opposite results, the source emits entangled particles. If there is no correlation, the particles are not entangled.

10.4

1. $|11111\rangle$
3. There are two ways to make the superposition: start with an alive worm (left) or start with a dead worm (right) (Fig. 2).

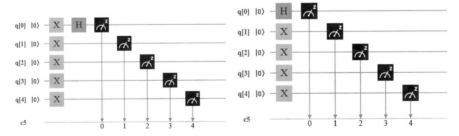

Fig. 2 Reprint courtesy of International Business Machines Corporation, © International Business Machines Corporation.

5. All quantum gates are reversible. Since a Hadamard gate made the superposition, its conjugate transpose can undo this. The Hadamard is its own conjugate transpose, so applying a second will undo the superposition.
7. $|11010\rangle$

9. 50% chance of alive and 50% chance of very dead

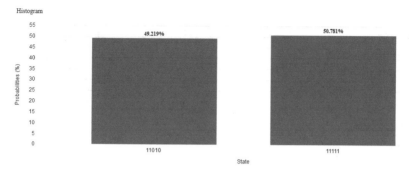

Fig. 3 Reprint courtesy of International Business Machines Corporation, © International Business Machines Corporation

10.5

1. C. The output is indistinguishable in the z-basis
3. Changing the superposition state from the z- to x-basis:

$$\frac{1}{\sqrt{2}}\left(|0\rangle + |1\rangle\right) = \frac{1}{\sqrt{2}}\left(\frac{1}{\sqrt{2}}|+\rangle + \frac{1}{\sqrt{2}}|-\rangle + \frac{1}{\sqrt{2}}|+\rangle - \frac{1}{\sqrt{2}}|-\rangle\right) = |+\rangle,$$

so only $+x$ will be measured. Whereas in a mix of $|0\rangle = \frac{1}{\sqrt{2}}|+\rangle + \frac{1}{\sqrt{2}}|-\rangle$ and $|1\rangle = \frac{1}{\sqrt{2}}|+\rangle - \frac{1}{\sqrt{2}}|-\rangle$, both $+x$ and $-x$ will be measured with 50/50 probability.
5. In the z-basis, we measure about 20% spin up and 80% spin down. Thus, $A = \frac{1}{5}$ and $B = \frac{4}{5}$.
7. Changing the superposition state from the z to x-basis:

$$\alpha|0\rangle + \beta|1\rangle = \frac{1}{\sqrt{5}}\left(\frac{1}{\sqrt{2}}|-\rangle\right) + \frac{2}{\sqrt{5}}\left(\frac{1}{\sqrt{2}}|+\rangle - \frac{1}{\sqrt{2}}|-\rangle\right) = \frac{3}{\sqrt{10}}|+\rangle - \frac{1}{\sqrt{10}}|-\rangle,$$

By squaring the amplitudes, we find 90% probability of $+x$ and 10% probability of $-x$.

10.6

3. Cosine squared function
7. Opposite of the probability for the spin up electron, so $1 - \cos^2(22.5°) = 0.146$.

Appendix A Solutions

A.2

1. (a) $\frac{5}{6}$, first die can be any number, second die can be any number but what the first die lands on.
   (b) Sum to greater than 8: $\frac{10}{36} = \frac{5}{18}$
       Sum to less than 8: $\frac{21}{36} = \frac{7}{12}$
   (c) Both even: $\frac{9}{36} = \frac{1}{4}$
       Both odd: $\frac{9}{36} = \frac{1}{4}$
       One even, one odd: $\frac{18}{36} = \frac{1}{2}$
3. (a) $\frac{1}{2} \times \frac{1}{3} = \frac{1}{6}$
   (b) (Face cards: Jack, Queen, King, Ace) $\frac{16}{52} \times \frac{1}{6} = \frac{2}{39}$
   (c) $\frac{1}{52} \times \frac{3}{6} = \frac{1}{104}$

A.4

1. Mean = 76.233
   Median = 73.5
   Mode = 72

A.5

1. Variance = 140.25
   Standard deviation = 11.843

Appendix B Solutions

1. $1 \times (-3) + 4 \times 2 + 3 \times 5 = 18$
3. $\begin{pmatrix} i & 0 \\ 0 & -i \end{pmatrix}$

Index

© The Author(s) 2021
C. Hughes et al., *Quantum Computing for the Quantum Curious*,
https://doi.org/10.1007/978-3-030-61601-4